The **NO-NONSENSE GUIDE** to

CLIMATE

THE SCIENCE, THE SOLUTIONS,

THE WAY FORWARD

'Publishers have created lists of short books that discuss the questions that your average [electoral] candidate will only ever touch if armed with a slogan and a soundbite. Together [such books] hint at a resurgence of the grand educational tradition... Closest to the hot headline issues are *The No-Nonsense Guides*. These target those topics that a large army of voters care about, but that politicos evade. Arguments, figures and documents combine to prove that good journalism is far too important to be left to (most) journalists.'

Boyd Tonkin,
The Independent,
London

WITHDRAWN

About the author
Danny Chivers is an environmental writer, researcher, professional carbon footprint analyst, activist and performance poet. He holds a BSc in Environmental Biology, an MSc in Nature, Science and Environmental Policy and an MProf in Leadership for Sustainable Development. He has carried out studies into the climate impact of electrical product manufacture, agriculture, food processing, international development NGOs, local authority carbon monitoring, retailers, offices and the UK government. Most recently, he has created an interactive emissions model of the UK economy for *The Guardian* website, co-founded 'Cyc du Soleil' (Britain's first mobile solar-and-cycle-powered performance stage), taken part in the Climate Camps at Heathrow, Kingsnorth and central London, and reached the semi-finals of the BBC Radio 4 National Poetry Slam.

Acknowledgements
I'm very grateful to everyone who helped and supported me through the writing process – the friends who commented on early drafts, our 'expert reader' Dr Chris Jardine, Nicola Bullard for the Foreword, my editor Chris Brazier for allowing me a few extra pages to explain the science properly while also keeping in the jokes, and Jess for being endlessly supportive despite my increasingly ridiculous working hours. You're all great.

About the New Internationalist
New Internationalist is an independent, not-for-profit publishing co-operative that reports on issues of global justice. We publish informative current affairs and popular reference titles, complemented by multicultural recipe books, photography and fiction from the Global South, as well as calendars, diaries and cards – all with a global justice world view.

If you like this *No-Nonsense Guide* you will also enjoy the **New Internationalist** magazine. The freshly designed magazine is packed full of quality writing, in-depth analysis and new features, including:
• Agenda: cutting-edge reports
• Argument: heated debate between experts
• Analysis: understanding the key global issues
• Action: making change happen
• Alternative living: inspiring ideas
• Arts: the best of global culture.

To find out more about the **New Internationalist**, visit our website at
www.newint.org

The  to

CLIMATE CHANGE
THE SCIENCE, THE SOLUTIONS, THE WAY FORWARD

Danny Chivers

NewInternationalist

The No-Nonsense Guide to Climate Change: The Science, the Solutions, the Way Forward.
Published in the UK in 2010 by New Internationalist™ Publications Ltd
55 Rectory Road
Oxford OX4 1BW, UK
www.newint.org
New Internationalist is a registered trade mark.

Cover image: Ho New/Reuters.

Series editor: Chris Brazier
Design by New Internationalist Publications Ltd.

Printed in UK by Bell and Bain Ltd
who hold environmental accreditation ISO 14001.

Mixed Sources
Product group from well-managed
forests and other controlled sources
www.fsc.org Cert no. TT-COC-002769
© 1996 Forest Stewardship Council

British Library Cataloguing-in-Publication Data.
A catalogue record for this book is available from the British Library.

Library of Congress Cataloguing-in-Publication Data.
A catalogue for this book is available from the Library of Congress.

ISBN 978-1-906523-85-5

Foreword

Most books on climate change focus on the science, usually getting so caught up in the complexities and catastrophes that they lose sight of the fact that climate change is the result of economic and energy systems based in societies, politics and power.

In this small book, Danny Chivers goes beyond the scary stuff and gets to grips with some big and complex ideas about climate change, science, economics, power, politics and history. In fact, he pins them down and makes them squirm, using wit, outrageous metaphors, lots of numbers and good, plain language. What's more, he does it wearing his heart on his sleeve, unashamedly arguing for justice over expediency, and with an optimist's eye, showing us the many ways in which life could actually be better for almost everyone if we do the right thing.

Many of the ideas and analyses that Danny rattles through at a clipping pace are known – at least at an intuitive level – to most climate justice activists: that economic growth based on the destruction of forests, land, rivers and ecosystems is a catastrophe for people and the planet; that corporate greed and entrenched power are major obstacles to action to halt climate change; that false ideas of progress founded on the endless consumption and production of 'stuff' is not the same as happiness and living well.

Fishers and farmers facing the Titans of agribusiness, communities ousted by mining companies, and indigenous peoples whose forests have been destroyed by logging and palm-oil plantations: all of these need no convincing that the system is broken and needs fixing.

But what Danny is trying to do – and in this I sincerely hope he succeeds for the sake of us all – is to convince comfortable consumers in the North that we have nothing to lose but our chains, so to speak. He challenges us to re-imagine the meaning of life without a 56-inch plasma television screen and a private car. He knows, though, that it's more than just a question of composting and bicycles: in the poem 'Lifestyle Choice' (page 124) he amusingly mocks those who believe that cotton shopping bags and low-energy light bulbs can do the trick. What if, he asks

... *the abolitionists, instead of fighting slavery,*

Just stayed at home and put a bit less sugar in their tea.

The urgent message of this book is not the gathering science showing that things are much worse than we thought (even though that's pretty scary) but the compelling argument that the only way we're going to get out of the fine mess we are in, is by building a vast, diverse and radical movement for climate justice, joining together frontline communities in the North and in the South to change the system. And, to change the system, we have to start by turning off the fossil-fuel tap. As Danny says:

All the renewable technology in the world won't help us if we're still digging up fossil fuels and ripping down the rainforests. We need to tackle both ends of this problem, and find ways to keep the oil in the soil, the coal in the hole, the gas in the crevasse and the trees...er...swaying gently in the breeze.

In this small, excellent book, we have plenty of ammunition to take on the corporate lobbyists, the climate deniers, the carbon traders and the nay-sayers.

Nicola Bullard,
Focus on the Global South, Bangkok, Thailand

CONTENTS

Introduction

THANKS FOR PICKING up this book. You may be here because:

1. You want to know more about climate change (or need to know more, for your work or studies);
2. You don't want to read a big book on climate change;
3. This is a small book.

Welcome – this could well be the book for you. The aim of this little guide is to provide an overview of all the most important elements of climate change – the science, the politics, the economics, the solutions, and the possible ways forward. It is concise and accessible, but also offers a host of references and pointers that can lead you to more detailed information. So if you think you already know about climate change but want a pocket guide for looking things up from time to time, this could also be the book you're after.

Some books claim to provide a balanced, unbiased view of the issue of climate change. This is a nice idea, but impossible. Climate change is something that affects, and is affected by, almost everything in our lives. There's no way for anyone to write about it without their own perspectives, beliefs and preferences creeping in.

So I'm going to be honest and tell you upfront where I'm coming from. These are my starting points:

• Climate change is a serious issue and will affect, or is already affecting, everyone (if you're not yet convinced about this, Chapter 1 will explain in more detail).

• There are many other vital issues in the world apart from climate change, such as poverty, inequality, war and oppression – or just the day-to-day struggle to get by that so many people face. It's important that the solutions to climate change don't make these problems worse. In fact, we should really try to find ways to tackle climate change that help with these other problems at the same time.

• The solutions to climate change should be as fair as possible, with the people who are most responsible for the problem having to put the most time, energy and money into the solutions. The people most affected by climate change should also have a big say in how to solve it. This just seems like basic fairness, really.

• Climate change isn't just a technical issue to do with putting the wrong amount of certain gases into the air. It's tangled up with politics, lifestyles, economics, power structures, culture and belief. This is why it's proving so difficult to solve, and also why it's simultaneously disastrous, frustrating, fascinating, heart-breaking, and utterly relevant to everyone in the world.

I hope that all sounds reasonable. If it doesn't – if you think that everything's just hunky-dory in the world, or that climate change is just a side issue that hasn't got much to do with anything else – I urge you to read this book anyway. Let me try to persuade you. It'll be fun.

A bit more about me, for the record: I carry out research into climate change for a living, and also spend a lot of time campaigning, speaking and performing on the topic. I don't earn very much money from this and probably never will.

This book is published by New Internationalist, which also produces an excellent magazine on international issues and the ongoing struggle for global justice. It's well worth checking out for the latest news and analysis on climate change (and other global issues) – www.newint.org.

OK – I think that's everything. Let's go.

Danny Chivers
Oxford

Part A: The Science

1 How do we know that climate change is happening?

The science behind climate change, clear and simple... How the greenhouse effect works... How temperatures are rising... Droughts, floods and storms... And how all this fits together.

LET ME TELL YOU a secret. Sometimes – just sometimes – I get jealous of the people who don't believe in climate change.

On those days, when I hear someone on breakfast radio declaring they have 'proof' that climate change isn't real, I give a cry of joy, leap out of bed and eagerly start investigating this wonderful claim, only to find that – as usual – they're talking absolute nonsense and the science of climate change is as frustratingly solid as ever.

I hope you don't blame me for thinking like this. The science isn't just solid, it's also pretty darned scary. The idea that the fuels that heat and power our lives – oil, coal and gas – are causing disastrous floods, storms and droughts all around the world is a highly disturbing one. It's no surprise, then, that 2010 has seen a rise in the number of people in industrialized nations who say they don't think climate change is a real problem – or that it somehow isn't humanity's fault.* As the science has become scarier, the siren voices of the professional climate change deniers – some funded by the fossil-fuel industry, others just

---

* In a UK poll by Ipsos Mori in February 2010, only 31% of people said climate change was 'definitely' happening. This was a drop from 44% the year before. This isn't quite as negative a result as it sounds – in answer to the same 2010 question, 29% said climate change 'was looking like a reality', 31% thought it was 'exaggerated', but only 6% said it wasn't happening at all (3% answered 'don't know'). A US poll by *Washington Post*-ABC News in November 2009 was more stark in its results: 26% thought that global temperatures were not rising, compared with 18% the year before.

basking in the limelight of controversy – have become harder and harder to resist.

It doesn't have to be like this, of course – there is an upside (sort of) to climate change, which is that if we pick the right solutions we could actually make our lives better in many ways, and help to make the world a fairer place. If we can move away from being scared of climate change, and start talking about the benefits of well-built homes, decent transport systems, healthy fresh food, cleaner local energy, and a fairer sharing of the world's land and resources, then it's far more likely that people will be active and enthused about climate change and less likely that they will blame it all on invisible sunspots or secret lizard conspiracies.

But before we get to all this, we have to be absolutely clear about the science. This chapter aims to give you a basic grounding in what climate change/global warming is, and how it works. It won't take long or get too technical, I promise – I'm assuming that like most people (including me), you don't have a degree in atmospheric chemistry. Luckily, we don't need one – the fundamental science behind climate change is pretty straightforward and easy to understand. There are five key points that, taken together, show us that climate change is both real and serious:
1) Carbon dioxide is a greenhouse gas
2) We've put loads of carbon dioxide (and other greenhouse gases) into the atmosphere
3) The average temperature of the planet has been rising
4) We've seen lots of other climate change effects
5) All of these things are connected.

The rest of this chapter examines each of these points in turn, explains the evidence for them and why they're significant. Each of the five points is accompanied by a 'Skeptics' Corner' box, where I'll present some common fallacies about climate change and explain what's wrong with them.

1) Carbon dioxide is a greenhouse gas

Nineteenth-century greenhouses

Back in the 1800s, a number of scientists were mucking about with gases in order to learn more about how the atmosphere worked. The French mathematician Joseph Fourier had realized in the 1820s that there must be something in the air that prevented the Sun's heat from just bouncing off the Earth and vanishing back into space.[1] In the 1860s, the Irish-born physicist John Tyndall experimented with a number of gases to see which were best at trapping heat – and found that carbon dioxide had the intriguing property of letting visible light pass through, but hanging on to heat.[2] His work was taken further by the Swedish Nobel Prize winner Svante Arrhenius in 1896, who linked the amount of carbon dioxide in the air to changes in global temperature.[3]

So the basic science behind climate change is nothing new – but over the last hundred years, scientists have collected a huge amount of evidence so as better to understand and document this phenomenon. Light from the Sun passes through the atmosphere, bounces off the Earth and heads back towards space. Carbon dioxide, water vapor, methane, and other heat-trapping gases hold back some of that reflected energy as heat, and thus the atmosphere – and the planet – warms up. Seemingly small changes in the levels of these 'greenhouse gases'* can lead to large changes in the Earth's temperature. Prehistoric records (see the next section) show that a shift in greenhouse gas levels from 0.02 per cent of the atmosphere to 0.03 per cent (from 200 to 300 parts per million) can be the difference between an ice age and what we think of as a 'normal' climate.**

---

* So-called because they reminded early scientists of the glass in a greenhouse.
** The long-term cooling and warming of the Earth over millions of years has been caused by a number of factors (such as fluctuations in the Earth's orbit and gradual geological shifts), not just greenhouse gas levels. However, greenhouse gases such as carbon dioxide have played a major role in speeding up and augmenting these ancient periods of warming and cooling – see Section 2, below.

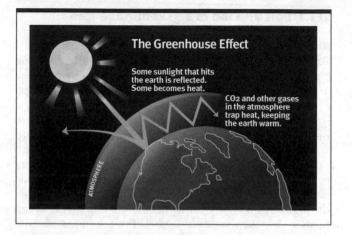

The joys of moderation

This is all well-established, non-controversial stuff. Without the greenhouse effect, the Earth would be a frozen lump of rock like the Moon. At the other end of the scale, Venus's atmosphere is 96-per-cent carbon dioxide, which, combined with its proximity to the Sun, gives it a balmy surface temperature of around 480°C.

Water is the most abundant greenhouse gas in the atmosphere, and makes the biggest contribution to the 'natural' greenhouse effect. However, the amount of water vapor in the atmosphere stays pretty much the same from year to year, so it doesn't play much of a role in the story of modern climate change (or at least, not yet – see Chapter 2). The second most common greenhouse gas is carbon dioxide – CO_2 for short. This is the one we really need to know about. Even though there's less CO_2 than water in the atmosphere, it's still the most important greenhouse gas as far as today's climate change is concerned, because – unlike water vapor – the CO_2 in the atmosphere is increasing rapidly (see point 2, below), and it stays in the air for a long time (around 200 years). There are several other important gases which also have a warming effect, as shown in the table below.

Key greenhouse gases affecting modern global warming

Greenhouse Gas	Heat absorbed over 100 years (compared to CO_2)	How much is in the atmosphere (in parts per million)
Carbon Dioxide (CO_2)	1	388
Methane (CH_4)	A kg of methane traps 25 times more heat than a kg of CO_2	1.8
Nitrous Oxide (N_2O)	A kg of nitrous oxide traps 298 times more heat than a kg of CO_2	0.3
Artificial industrial gases hydrofluorocarbons (HFCs), perfluorocarbons (PFCs), and sulfur hexafluoride (SF6)	A kg of one of these gases can trap between 140 and 24,000 times more heat than a kg of CO_2	Less than 0.001

Source: IPCC Fourth Assessment Report, 2007.

In the right quantities, greenhouse gases are crucial for keeping the planet within a temperature range that allows life as we know it to survive.

Skeptics' corner: CO_2 as a greenhouse gas
This is basic, well-established science that is very difficult to deny. In fact, you can demonstrate it yourself by filling a plastic bottle with carbon dioxide, shining a lamp on it and measuring its temperature.[4] ∎

2) We have put loads of extra greenhouse gases into the atmosphere

Climbing the sawblade
We now fast-forward from the 1890s to the 1950s, and a young American chemist called Charles Keeling. With no greater aim in mind than setting himself an interesting challenge, he worked out a more

accurate way of measuring the quantity of carbon dioxide in the atmosphere. In 1958, the US Weather Bureau started using Keeling's new technique at their monitoring station on Mauna Loa in Hawaii – and were surprised to find that CO_2 levels in the Earth's atmosphere were increasing at a significant rate, year-on-year.[5] These measurements have been taken ever since, and form the graph below (also known as the 'Keeling Curve').

The sharp little oscillations in the graph are caused by the great forests of the Northern hemisphere; they take in more CO_2 during summer, and release more during winter, turning the graph from a smooth curve into a sawblade.[6] But the trend is clearly an upward one – and it's also accelerating. If you compare the left-hand to the right-hand end of the line, you'll see that the increase has grown gradually sharper over the last 40 years – we've been churning out the CO_2 faster and faster as time's gone on.

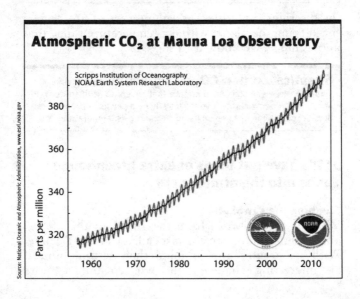

Going with the flow

But how do we know that this increase is all the fault of humanity? Mightn't this CO_2 be coming from somewhere else? It's a reasonable question, and in order to answer it scientists have built up a detailed picture of global carbon flows. The table below shows where all the carbon on the planet is, and how it is currently moving around. At the moment, most of this carbon isn't in the form of CO_2 – it's part of the rocks or plants or fossil fuels. To make it easier to compare with everything else in this book, I've converted all the carbon in this table to carbon dioxide equivalent

Global carbon cycle in billions of tonnes of CO_2 equivalent

Location	How much carbon is stored there	Amount released to air per year	Amount absorbed from the air/year	Net release to atmosphere /year
Earth's Crust	370,000,000	0.4 (from volcanoes)	0.4 (long-term absorption by rocks)	0
Oceans	141,000	332 (released to the air)	340 (dissolved from the air and washed in by rivers)	-8
Fossil Fuels	14,000	28 (from humanity burning the stuff)	0	28
Plants and Soils	8,500	444 (natural plant respiration plus deforestation, peat cutting etc)	449 (photosynthesis and afforestation)	-5
TOTAL	370,163,500	804	789	15

Source: UNESCO/SCOPE/UNEP http://nin.tl/bc7rhm Gigatonnes of carbon converted into billions of tonnes of CO_2e by my own calculation (multiplied by 3.67). All figures are from UNESCO, for 2008. More recent (2010) annual emissions from the burning of fossil fuel were closer to 31 billion tonnes/year. CO_2 from cement manufacture (see below) is not included. Note: while the 'amount stored' figures are approximate, the amounts of carbon moving in and out have been studied in detail.

(CO_2e) – in other words, how much CO_2 would be produced if all this carbon was burned and released into the atmosphere.

From this table, you can see that while CO_2 is being released from natural sources – oceans, plants, soils and rocks – these natural carbon stores are in fact sucking up slightly more than they are putting out each year. However, the burning of fossil fuels by humanity has shoved a spanner in the spokes of the carbon cycle, by releasing more CO_2 per year than these natural systems can absorb – which is why this key greenhouse gas is building up in the atmosphere, and why the Mauna Loa measurement is a little bit higher every year.

To be totally certain that all this extra CO_2 is coming from humans, there are two other things we can check. First, we can look at historical records of how much fossil fuel humanity has burned each year, for the last 250 years or so. We know how much CO_2 is emitted for each kg of coal, oil or gas that we burn, which means we can make a graph of humanity's approximate CO_2 emissions stretching back to 1750.

This tallies neatly with the rising CO_2 levels in the atmosphere.

Finally, just to check that humanity really is the main source, and that there isn't some giant

---

Fossil fuels

In prehistoric times, CO_2 levels gradually rose and fell as the gas was absorbed or released by plants, rocks and oceans in response to various natural cycles. Over time, billions of tonnes of CO_2 were removed from the air by the great forests of the carboniferous period, and the carbon was stored underground as coal. Billions more tonnes of carbon were sealed away as oil and natural gas, formed from fossilized sea creatures. When we burn these fuels to power our hedge-trimmers, latte makers and shampoo factories, the carbon combines with oxygen, and CO_2 is released back into the air. ■

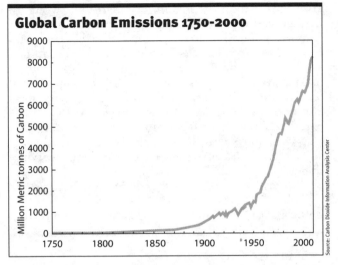

Global Carbon Emissions 1750-2000

Million Metric tonnes of Carbon

9000 — 8000 — 7000 — 6000 — 5000 — 4000 — 3000 — 2000 — 1000 — 0

1750   1800   1850   1900   1950   2000

Source: Carbon Dioxide Information Analysis Center

underwater CO_2-spewing volcano the size of Australia that no-one's discovered yet, we can look at something called the *isotopic signature* of the carbon dioxide in the air. Without wanting to get too technical, carbon from different sources has some very slight physical differences that we can use to figure out where it's from. Scientists started checking up on this in the early 1980s, and found that most of the extra carbon appearing in the atmosphere was the type of carbon that comes from fossil fuels.[7]

At the time of writing (August 2010), the average amount of CO_2 in the atmosphere stands at 388 parts per million (ppm).[8] But 'parts per million' doesn't sound like very much: is that really a high enough level to have a noticeable effect on the world's climate? We can start to answer this question with a bit of time travel.

Cold hard facts

Between 1990 and 1998, a narrow but very deep (3.5-kilometer) column was drilled from the ice

How do we know that climate change is happening?

at Vostok, Antarctica. This 'ice core' is a piece of frozen history – it contains many tiny bubbles of trapped air dating back through the last 400,000 years. Each of these is a miniature time capsule, and can be analyzed to tell us what gases were in the atmosphere as well as the average global temperature at that point in (pre)history.[9] This allows us to stretch our Keeling graph of CO_2 levels back in time a few eons, and see how current CO_2 levels compare with the past (below).

This graph shows us three interesting (and worrying) things:

1. The Earth's temperature clearly falls and rises in line with CO_2 levels in the atmosphere.
2. In the last 400,000 years, shifts in CO_2 levels from

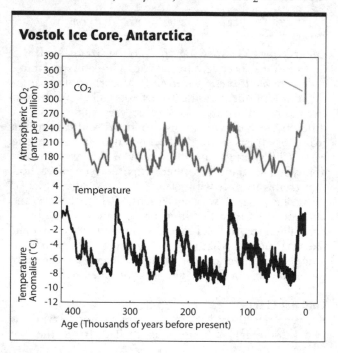

Vostok Ice Core, Antarctica

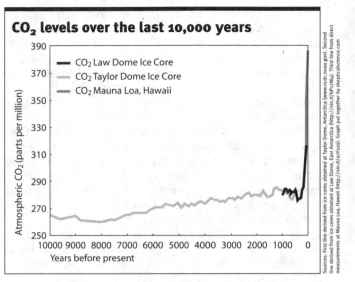

CO₂ levels over the last 10,000 years

Atmospheric CO_2 (parts per million)

- CO_2 Law Dome Ice Core
- CO_2 Taylor Dome Ice Core
- CO_2 Mauna Loa, Hawaii

Years before present

Sources: First line derived from ice cores obtained at Taylor Dome, Antarctica (www.ncdc.noaa.gov). Second line derived from ice cores obtained at Law Dome, East Antarctica (http://nin.tl/bP7yM4). Third line from direct measurements at Mauna Loa, Hawaii (http://nin.tl/a7Fsz0). Graph put together by skepticalscience.com

around 180 ppm to 300 ppm have been associated with very serious changes in climate – the low points on the temperature graph correspond with what we think of as Ice Ages.*

3. We are currently at 388 parts per million of CO_2, right at the top of the graph. This is higher than at any other point in the last 400,000 years. Also, the current rate of increase is so fast that it looks like a vertical line on this graph.

Unsurprisingly, CO_2 levels began rising at the beginning of the Industrial Revolution (the late 18th century), when humanity began to burn fossil fuels in earnest and have gradually accelerated ever since. If we 'zoom in' on the last 10,000 years of CO_2 data from more recent ice cores, we can see this quite clearly (see graph above).

---

* The Vostok Ice Core graph shows changes in Antarctic temperature of up to 12°C. Temperature changes elsewhere on the planet were slightly less drastic than this (though still large enough to cause Ice Ages), because temperatures tend to change further and faster at the planet's poles.

How do we know that climate change is happening?

The last time CO_2 is believed to have been this high was the mid-Pliocene period, some 3.5 million years ago. At this time, sea levels were 25 meters higher than today.[10] Clearly, the changes we are making to atmospheric CO_2 are in the 'significant' range – but then, if CO_2 is at such a (pre)historic high, then why isn't Europe under water yet? Why isn't Greenland famous for its tropical beach holidays?

The answer is that – fortunately – there's a time delay between rising CO_2 and rising atmospheric temperatures.[11] The greenhouse gases need to spend a decent amount of time in the atmosphere, doing their warming thing, before we start to really notice the effects down on Earth. This is because the oceans, rather than the air and the land, are absorbing most of the extra heat. The effects of this oceanic warming are far less immediate and noticeable to us land-dwellers than the effects of a warmer atmosphere. This is something of a double-edged sword. On the plus side, this delayed reaction means that the effects of climate change have been nowhere near as extreme as they would have been if the ocean wasn't such a great heat trap. However, on the negative side, it means that it's much harder for the person on the street (or field, or hillside) to see the connection between our CO_2 emissions and the gradual warming of the planet.

Emission omissions

More about this a bit later, but first some more bad news (sorry). So far, we've mostly been discussing the CO_2 released from the burning of fossil fuels. However, this isn't the only way in which humanity is contributing to climate change:

- Cement manufacture brings some extra CO_2 into the equation, because it involves heating calcium carbonate (from rocks) to produce lime and carbon dioxide – in other words, taking carbon from the

'Earth's Crust' store in the table on page 17 and putting it into the air. This releases an extra 0.9 billion tonnes of CO_2 into the atmosphere each year (about three per cent of the annual CO_2 total).[12]

- We're releasing greenhouse gases other than CO_2 into the air. The most important ones are methane (from livestock, coal mines, rice fields and landfill sites), nitrous oxide (from fertilizers and manufacturing) and small amounts of other powerful warming gases from certain industrial processes. Together, these bonus greenhouse gases add an extra 30 per cent to the warming power of our annual global CO_2 emissions.[13]

- The emissions from airplanes cause a bit of extra warming, because they occur high up in the atmosphere (there's a more detailed explanation of

Skeptics' corner: Rising CO_2 levels

The fact that CO_2 levels are increasing is pretty undeniable, because they're directly measured. Even though they're measured in parts per million, these relatively small amounts of greenhouse gas are extremely significant – without them, the Earth would be a lump of frozen rock, and none of us would ever have been born. We've also seen how relatively small changes in CO_2 levels have, in the past, led to major changes in the planet's climate.

Some commentators dispute this latter point – they note that, judging by the prehistoric ice core record, when the Earth has warmed in the past the temperature has tended to rise before the CO_2, suggesting that the warming was causing the CO_2 increase rather than the other way around. I'll be honest – if you're not familiar with the science, this can be pretty confusing. In reality, it's a well-established part of climate science – see the 'Lessons from Prehistory' box for a full explanation.

Some prominent 'skeptics' such as Ian Plimer have made the bizarre claim that volcanoes produce more CO_2 than all of humanity's activities.[14] Referring back to the table on page 17, we can see that this clearly isn't true – humanity is responsible for at least 60 times the CO_2 of volcanoes. This point was firmly underlined when the brilliantly named Icelandic volcano Eyjafjallajokull erupted in March 2010; the resulting ash cloud led to the grounding of planes across Europe for several weeks, preventing the release of far more carbon dioxide than the volcano was emitting. It created a net saving of around 50,000 tonnes of CO_2 per day.[15] ∎

Lessons from prehistory: ancient warming cycles and what they can tell us

Scientists are developing more and more powerful computer models to try to predict the effects of climate change, but there's one model that will always beat them all – the planet Earth itself. By looking at what's happened when CO_2 levels have risen in the past, we can learn a huge amount about what's likely to happen in response to our current CO_2-emitting frenzy.

We can gather evidence from ancient ice cores, tree rings, coastlines, and the ocean's depths that provide us with a pretty decent picture of how temperatures, sea levels, and the amount of CO_2 in the air have changed over the last few tens of millions of years. The results are fascinating – the Earth has swung periodically between colder and warmer periods over the eons (for example, see the the graph on page 20). In the coldest periods ('glaciations' or ice ages), the Northern continents were covered with massive ice sheets several kilometers thick; in the warmer periods, there was no ice at the poles and sea levels were up to 75 meters higher than today. These huge changes were initially triggered by tiny fluctuations in the Earth's temperature – the Sun might go through a slightly warmer or cooler phase,[17] or kinks in the Earth's orbit might take the planet out a little further or in a little closer.[18] This would result in an incredibly small amount of extra warming or cooling each year. The planet would then warm or cool gradually, over hundreds or thousands of years. Then suddenly, this would transform into rapid change, switching the planet from cool to warm or vice versa.

Why the sudden flip into rapid change? Well, this is where we need to start talking about 'feedbacks' – factors which can either speed up or slow down the rate of global warming or cooling. There are three particularly important feedbacks when it comes to prehistoric warming:

1) Carbon dioxide and methane release: As the planet warms up, carbon dioxide and methane are released from plants, soils and oceans. These gases create a greenhouse effect which leads to more warming and thus the release of more CO_2 and so on until the whole climate has changed completely. This has been a very important factor in transforming very slow prehistoric warmings into sudden shifts. This explains why, in the

this in Chapter 4).

- Finally – and vitally – we are messing with the planet's ability to suck up carbon dioxide. As we saw in the table on page 17, about 789 billion tonnes of CO_2 are absorbed by plants, soils and oceans each year. But we are in the process of cutting down forests faster than at any other point graph on page 20, temperatures start to rise first, and then CO_2 follows – in each of the roughly 5,000-year warming periods shown by this graph, other factors accounted for about the first 800 years of warming; the following 4,200 years were then due mainly to CO_2 levels.

2) Ice cover: As things gradually get hotter, snow and ice start to melt at the poles. The white reflective surface gives way to reveal the much darker land or water underneath. This absorbs more heat from the sun, thus speeding up the warming of the planet. This is called the 'albedo effect' (albedo is the technical term for 'reflectiveness').

3) Water vapor: warmer temperatures increase evaporation, putting more water vapor into the air. You may remember from the beginning of this chapter that water vapor is a greenhouse gas, so this process also adds to global warming.

All three of these processes also work in reverse – if the Sun or the Earth's orbit shift to a slightly colder phase, there will be a period of slow cooling until falling CO_2 levels, increased ice cover and/or a drop in water vapor cause the temperature to plunge.

Eventually, the warming or cooling period will reach its natural end, when the feedbacks run out of power (for example, if the air can't hold any more water vapor, the maximum amount of CO_2 has been released, and all the ice has melted), or when external solar or orbital changes start to push things back the other way. Sometimes the Earth will remain in a relatively stable state for hundreds, thousands or millions of years, as all these forces come into balance; at other times the temperature will slowly start to change again, in response to the next tiny fluctuation in the sun's output or in our planet's kinky orbit.[19]

By pumping such a large amount of CO_2 into the air at an unprecedented rate (see the right-hand end of the graphs on pages 20 and 21), we have effectively overridden these slow natural cycles. The jolt of extra heating from humanity's greenhouse gas emissions is happening much more quickly than the tiny, incremental changes that triggered those past warmings. We now have to hope that we can reduce CO_2 levels rapidly enough to prevent those same feedback mechanisms kicking in that have changed the Earth's climate so drastically in the past (see Chapter 2). ■

in history – which is a double climate whammy, because it releases the carbon stored in those trees (and soil) back into the air, and also means there are fewer trees around to suck up CO_2 from that point onwards.[16] This is Not a Good Thing.

There's more information on all of these emissions sources in Chapter 4. For now, it's enough to say that yes, we have definitely added greenhouse gases to the atmosphere, and in large enough quantities to expect to see an impact on the Earth's climate.

3) The average temperature of the planet has been rising

Hot news

It's August 2010, and I'm looking at a series of newspaper headlines from the last few weeks:

'World feeling the heat as 17 countries experience record temperatures' – *The Guardian*[20]

'Long, hot summer of fire, floods fits predictions' – Associated Press[21]

'Warming world records hottest ever June' – *The Australian*[22]

'This year warmest on record so far' – *The Washington Post*[23]

I can't help having mixed feelings about this. Rising global temperatures are seriously bad news, bringing all sorts of awful effects for people and the natural world (see below). But maybe, just maybe, this year of record temperatures will be the extra shove we need to move enough residents of industrialized nations from dismissal and denial into serious climate action.

Well, let's hope so (but see Chapter 3 for some reasons why it isn't quite that simple). In the meantime, 2010's super-scorcher summer adds another sizeable boulder onto the mountain of evidence for a rapidly warming world:

• The global average temperature has risen by 0.8 degrees since pre-industrial times.[24] This brings us up to an average temperature higher than at any time in the last 100,000 years.[25]

• The rate of warming has been increasing further in the last few decades. The warming trend for the last 25 years is more than double the rate of the previous 100 years, and the 10 hottest years on record have all occurred since 1990.[26]

• Extreme temperature events are also becoming more common – for example, the 2003 European heatwave shattered records and caused 35,000 deaths.

• The winter of 2006/07 was the warmest across the Northern Hemisphere since records began.[27] The world downhill ski championships in Austria were thrown into crisis by a lack of snow[28] and Spanish bears gave up on hibernation.[29] In the UK, spring flowers popped up in January – and came up even earlier the following winter.[30]

• Since 2005, Arctic sea ice has been melting faster than scientists' predictions;[31] the Greenland ice shelf's summer melt period has increased by 16% over the last 30 years[32] and 2010 has seen the fastest Greenland melt ever recorded.[33]

The heat is on
All of these examples fit into a broader trend, as shown in the next three graphs (below and overleaf).

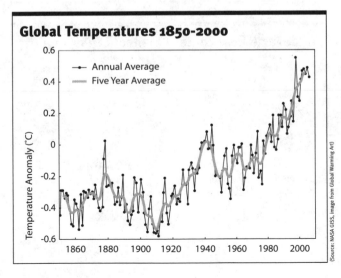

(Source: NASA GISS, image from Global Warming Art)

How do we know that climate change is happening?

Sources: Mann, Bradley and Hughes (1999) 'Northern Hemisphere Temperatures During the Past Millennium: Inferences, Uncertainties, and Limitations', Geophysical Research Letters, 26, 6, 1999; and IPCC Third Assessment Report, Working Group 1, Figure 2.20.

Global temperatures over the last 1000 years

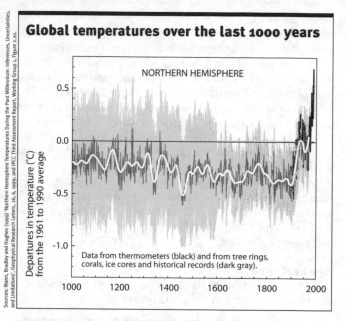

NORTHERN HEMISPHERE

Departures in temperature (°C) from the 1961 to 1990 average

Data from thermometers (black) and from tree rings, corals, ice cores and historical records (dark gray).

Sources: Murphy et al., 'An observationally based energy balance for the Earth since 1950', Journal of Geophysical Research, 114, 2009; Domingues et al., 'Improved estimates of upper-ocean warming and multi-decadal sea-level rise', Nature, 453, 2008; www.skepticalscience.com

Total Earth heat content anomaly from 1950

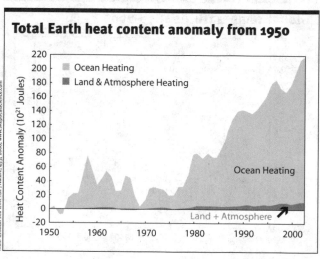

Heat Content Anomaly (10²¹ Joules)

Ocean Heating
Land & Atmosphere Heating

Ocean Heating

Land + Atmosphere

The figures in the top graph opposite show average temperature measurements for the land and the air. However, it's becoming increasingly clear that these temperature changes are being dwarfed by rising ocean temperatures. The bottom graph shows the latest estimates for how changes in ocean temperatures fit into the picture (corresponding with the final 60 years of the top graph).

How do we know that temperatures are rising?

Taking the Earth's temperature is a bit of a challenge, because the planet is just so darned big. Anywhere we stick a thermometer will just give us the temperature of that particular location. To get a meaningful global average, we need rather more data than that.

Fortunately, over the last few centuries, local weather agencies have gradually built up a huge network of measurement stations all across the globe. Today, there are more than 7,000 of them, over both land and sea, and their results are collected and compiled by three different organizations.[34] This means that any strange results or anomalies can be checked against a different set of data, and any mistakes are easy to spot. Satellites orbiting the Earth are also increasingly being used for thermal imaging, supplying us with yet more temperature data.[35] There are, inevitably, slight differences between the results from different sources (due to instrumental errors, calculation methods etc) but the overall trend is very clear – the planet is warming. The graphs in this section have been compiled from all of these data sources.

The scientists doing the measuring would be the first to admit that achieving an accurate result isn't easy. Here are a couple of reasons why:
• Thermometer design and accuracy have changed over the last 150 years.
• The urban 'heat island effect' – put simply, towns and cities are warmer than the countryside, and most of the data collection points are located in the industrialized, urbanized world, which creates some slight biases in the data.

These factors are carefully taken into account when calculating global average temperatures. In early 2010, climate scientists were given some unintended help in this area when the notorious anti-climate-science blogger Anthony Watts launched a US-wide survey of temperature measurement stations, rating them 'good' or 'bad' based on their proximity to towns and other heat sources.[36] He was hoping to prove that the heat island effect was responsible for the rising temperatures on all those official US thermometers – but instead found precisely the opposite. Watts' results showed that the heat island effect was, in fact, having less of an impact than the scientists had thought, and so caused them to adjust the US temperature record slightly upwards rather than downwards.[37] Whoops. ■

Skeptics' Corner: A warming world

Weirdly, despite the huge body of evidence for rising planetary temperatures, people determined to deny the existence of climate change have focused a lot of their efforts on this topic lately. Here are some common misleading claims, with explanations.

Global warming stopped in 1998! Perhaps 2010's high temperatures will put paid to this one, but just for the record... The Climate Research Unit at the UK's University of East Anglia is one of the three organizations that compile surface temperature records. If you look at their data alone, it looks as though 1998 was slightly warmer than the years 1999-2009.[38] Some commentators have seized on this as evidence that global warming has 'stopped'. However, this is based only on land and air temperatures, and not the total picture – when you look at the land, air and ocean all together (as in the bottom graph on page 28) the warming trend is still clearly heading upwards. In any case, if you look at the average of all the temperature data from across the world (rather than just the CRU data), 2005 was hotter than 1998 overall.

The world cooled down between 1940 and 1970! This isn't true, but is based on something genuinely interesting. Global warming **did** plateau for a bit between the 1940s and the 1970s, due to the now widely recognized phenomenon of 'global dimming' – a type of industrial pollutant called sulfate aerosols were partially blocking the Sun's rays. This lasted for a while until the ongoing build-up of greenhouse gases – combined, ironically enough, with a reduction in sulfate pollution from power stations – eventually swamped the dimming effect and the temperature began to rise once more.[39] You can see this flattened period on the graph on page 27.

The temperature measurements are flawed/manipulated/fixed! To tamper with or subvert the data from 7,000 different measure-

So the temperature changes we are feeling on land are small fry (if you'll pardon the expression) compared to the heating of the oceans. The land and air may be getting sunburned, but the oceans are in the toaster. The full consequences of this remain to be seen.

That's probably enough for now – suffice to say that yes, the earth is definitely warming at a far greater rate than it has for a very long time.

ment stations and satellites, which are processed via three different major organizations with hundreds of staff, would require an utterly fantastic level of conspiracy which would put a James Bond baddie to shame. Nonetheless, in 2010, a procession of (mostly online) commentators claimed that a series of hacked emails from the Climate Research Unit at the University of East Anglia contained evidence of just such a conspiracy. Three separate independent enquiries trawled painstakingly through the emails and found evidence of nothing more than a few scientists occasionally being a bit rude about some of their colleagues, using some unhelpful jargon and having the odd moan about incessant public requests for information (pretty much like the emails from any other workplace).[40] However, there was one useful outcome from this: much more of the raw temperature data has since been made public, to avoid similar accusations in the future.[41]

But it's cold today...
Although the average global temperature is rising, that doesn't mean that everywhere is getting hotter at the same rate. The global climate system is complicated; some places are heating up faster than others, and some places may even cool down depending on how ocean currents shift around. There's also an important difference between *climate* and *weather*. Climate change is a gradual, long-term process; weather is about short-term, day-to-day changes due to local patterns of wind, evaporation, and ocean currents, and is much more chaotic and unpredictable. A few weeks of cold weather in one location tells us little about long-term global temperature change – that's why we need all those thousands of temperature measurement stations taking decades' worth of readings. Those measurements are telling us that every time there's a bit of unusually cold weather somewhere in the world, it's being outweighed by many more examples of unusually hot weather elsewhere, and so the overall trend is of a warming planet. ∎

4) We've been seeing lots of other effects

Warm words
There's a reason why campaigners tend not to use the term 'global warming' these days – not only does it sound a bit too pleasant (who in the industrialized North wouldn't want a nice bit of warming?), but also it doesn't describe the full effects that rising

global temperatures are having on the planet. 'Climate change' seems to have become the accepted term to sum up the bundle of different effects we can expect from an overheated planet. As campaigners such as George Marshall[42] have noted, these two words still sound rather innocuous – but more accurate descriptions such as 'climate carnage', 'weather frenzy' or 'total stormfest' probably sound a bit too dramatic to be taken seriously.

As I write this in 2010, Pakistan has just been hit by devastating floods, the worst in the country's history, killing around 1,600 people (so far) and leaving an estimated 20 million temporarily or permanently homeless. Russia is recovering from a blistering heatwave that saw temperatures consistently at 20°C above normal and caused thousands of deaths. Huge wildfires have been sweeping through northern Portugal, torrential rains have unleashed killer landslides in China, hundreds have been forced from their homes by floodwaters in Iowa, and Niger has just been hit with disastrous floods straight after a crippling drought. Reports like this flash past us all the time these days, and usually seem to be described as the worst in 20, 30, 50, 100 years.

No individual incident can ever be linked directly to climate change – the world's weather systems are far too complex for that. However, an overall trend of more frequent, and stronger, extreme weather events is exactly what we would expect from a warming world. It's a bit like cheating at dice. If you put a secret weight inside a die to make a six more likely to come up, then you'll roll more sixes; but you'll never be sure which of those extra sixes were due to the weight, and which would have come up anyway.

Cooking up a storm

When global temperatures rise, that doesn't just mean things getting hotter – it means that more and more energy is being pumped into the climate system. This has all sorts of knock-on effects,[43] including:

• More evaporation from ocean surfaces, leading to more powerful storms
• Greater heat transfer between different bits of the climate system, leading to stronger winds and more extreme temperatures (both high and low) in particular places
• Changes in rainfall patterns, leading to burst river banks in some areas and droughts in others
• Rising sea levels caused by the fact that warm water expands, as well as by the melting of Greenland and Antarctica.

This increase in weather-related disasters has been documented in detail by that well-known group of radical climate campaigners, large insurance companies. Over the page is a graph from the major insurers Munich Re, comparing the number of natural catastrophes per year from 1980 to 2009.[44] The trend is very clear, with more than twice as many disasters in 2009 as in 1980. Meanwhile, 2010 is on track to be as bad as 2009, with 440 disasters recorded between January and June.

It's also worth noting that the top three bars – all of which contain events which could be linked to

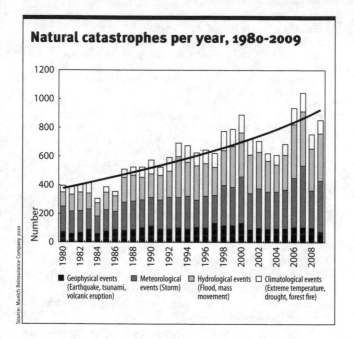

Natural catastrophes per year, 1980-2009

Source: Munich Reinsurance Company 2010

climate change – have grown noticeably. The bottom bar, which records natural disasters not linked to climate change (earthquakes and volcanic eruptions), has not grown in the same way.* The United Nations Environment Programme has found a similar trend stretching back to 1900.[45]

When interviewed on this topic in 2010, Liz Bentley – a climate scientist from the Royal Meteorological Society – said: 'Disasters such as the Boscastle flooding [in Cornwall in 2004] are moving from being a once-in-100-year event to a once-in-20-year event.'[46] Peter Stott, a climate scientist from Britain's Met Office, was asked about the 2003 heatwave in

---

* There is evidence that retreating ice sheets can increase the frequency of volcanic events in some areas, but this is currently limited to a very few places and so its global effect is small.

Europe, which killed around 35,000 people. He said the probability of these kinds of events had 'likely doubled as a result of human influence...[and] could become considered the norm by the middle of this century.'

The key word here is 'probability' – not every natural disaster will be the direct result of global warming, but we can expect the overall number of disasters to increase as climate change progresses. Recent disaster rates are exactly what we would expect from climate change at this point. According to the World Meteorological Organization, 'The sequence of current events matches [climate scientists'] projections of more frequent and more intense extreme weather events due to global warming.'[47]

These bare statistics don't really tell the story though, so here are some more specific examples, with a brief explanation of how each type of phenomenon may be linked to climate change.[48]

Storms from the sea

Hurricane-force winds can only arise from oceans with a surface temperature of at least 26°C. The warmer the oceans, the more intense and longer-lasting these storms are likely to be.[49] These kinds of tropical ocean storms are called hurricanes, typhoons or cyclones, depending on where you are in the world.

Hurricane Katrina hit the south coast of the US in 2005, causing 1,836 deaths. Later that year, Hurricane Stan hit Guatemala, Mexico, El Salvador, Nicaragua and Costa Rica, causing 1,500 further (but less globally reported) deaths. Some scientists argue that climate change contributed to the strength of these storms, though this is still hotly debated. 2004 saw the first South Atlantic hurricane ever observed hit Brazil, and in 2008 Cyclone Nargis blasted across Burma, killing an estimated 150,000 people and affecting over two million more.

How do we know that climate change is happening?

Floods from the sky

Higher temperatures mean more evaporation, which affects global rainfall patterns. As a result, some regions have become dryer and some have become wetter, with a significant increase in extreme rainfall events in the last few decades.[50]

Torrential rains triggered flash floods across Africa in 2007, affecting 22 countries, thousands of hectares of agricultural land and over a million people. Record rainfall in India in July 2005 killed almost 1,000 people; tens of thousands were displaced and hundreds killed by rain-driven floods and mudslides in Brazil in 2004. Record-breaking rains and severe flooding caused serious property damage, mass evacuations and loss of life in New Zealand in 2004 and Britain in 2007 and 2009.

Fields of dust

With increased evaporation, there's another side to the coin: while some places become wetter, others become drier. This has led to droughts becoming more common, especially in the tropics, since 1970; droughts have also become longer and more severe, and are affecting wider areas. Fires and famines are all-too-commonly the result.[51]

Australia has been hit by a series of severe droughts since 2002, with serious effects on agriculture and freshwater availability. In 2006, several consecutive years of drought left more than 17 million people starving in Djibouti, Ethiopia, Kenya and Somalia. 1998 saw one of Indonesia's worst-ever fire outbreaks, following the failure of the monsoon rains.

Cracks in the ice

The Arctic is one of the areas of the planet being hit first and hardest by climate change. Arctic temperatures are on a jagged upward trend – 2007 was the warmest year on record. This is leading to all sorts of changes

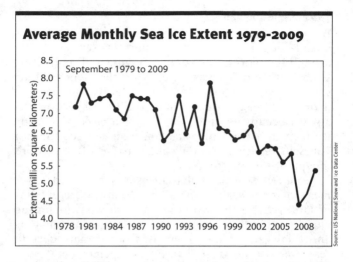

Average Monthly Sea Ice Extent 1979-2009

September 1979 to 2009

Extent (million square kilometers)

Source: US National Snow and Ice Data Center

to local wildlife movements and is threatening the traditional ways of life of the indigenous Arctic peoples.[52] Arctic sea ice grows and shrinks each year with the changing seasons, reaching its minimum size around September each year. Worryingly, this minimum extent has been on a gradual downward trend for the past 30 years (see graph above). Although – thankfully – the last two years on the graph weren't as bad as 2007, they still had less summer ice than any other year to date. Things are looking bad again for 2010. The speed of this thaw has shocked climate scientists: it is happening faster than the most extreme predictions made by the Intergovernmental Panel on Climate Change (IPCC).

Elsewhere in the Arctic, Greenland's ice has also been shrinking much more rapidly than expected.[53] Many observers were stunned in August 2010 when a huge chunk of ice, with an area of 160 square kilometers (four times the size of Manhattan), broke away from one of Greenland's main glaciers and tumbled into the sea.[54]

How do we know that climate change is happening?

The creeping oceans

As water gets warmer, it expands. This process has already led to a certain amount of sea level rise around the world, and is being augmented by melting Greenland and Antarctic ice. Note that when Arctic sea ice melts it does not make sea levels rise, because it produces just enough water to fill the space it was already occupying in the sea – watch a piece of floating ice melt in a glass of water and you can see this for yourself. Melting ice only adds to sea levels if it's moving from land to sea.

Sea level rise is difficult to measure – tides and waves mean that the sea doesn't stay still, and different bits of land around the world are also slowly moving up and down thanks to geological forces. Satellite imaging and tidal gauges have been used to estimate current sea level rise at around three millimeters per year.[55] This doesn't seem like much, but is already putting low-lying islands and coastlines at risk. Coastal farmers in Fiji are switching to salt-resistant crops, thanks to the surging ocean soaking their soils; meanwhile, the islands of Kiribati and Tuvalu are facing total evacuation as the sea creeps inexorably up through their towns and villages.[56]

Migrating from existence

As areas become warmer, dryer, or wetter, plants and animals are struggling to adapt. In the past, these kinds of changes in local climate have taken place over thousands of years, giving species time to evolve or to shift their range. The sheer pace of current change – combined with the fact that humanity's roads, towns, pollution and pipelines have seriously limited the number of places where these threatened species can go – is putting countless plants and animals at risk around the world.

The plight of the polar bears is well known – we can see their northern icy habitat melting and cracking beneath them, and so it's no surprise that several key populations in vulnerable areas are declining.[57] But any species which is adapted to a very specific set of temperature and/or weather conditions is likely to be affected by climate change, and most of them are less obvious and less well-studied than the polar bear. Our knowledge of exactly how our fellow species are faring is limited – it all depends on which ones we happen to be looking at closely enough.

For example, thanks to the enthusiasm of amateur butterfly-spotters past and present, we have enough data on British butterfly species to know that many of them are shifting their ranges northwards.[58] A lot of people keep an eye on migrating birds, and so we know that many of them are changing their migration times and routes in response to the changing climate.[59] Coral reefs are of huge interest to marine researchers (and holidaymakers), and so we've got plenty of information on how increased water temperatures are already killing off coral directly, or indirectly through the faster spread of coral diseases.[60]

But what about all the places where we're not looking, all the species we aren't keeping an eye on – or haven't even discovered yet? For example, around 35 species of barely studied Latin American frog are believed to have been wiped out by climate change,[61] including the extraordinary Golden Toad.[62] There's no way of telling how many other undiscovered species have gone the same way.

All life on Earth is connected, web-like; everything eats or is eaten by something, and competes with others for food, light or living space, so these changes in various organisms' range, numbers and behavior will have unpredictable knock-on effects for other species all around the world – including humans.

Life, limb and livelihood
As we've seen, climate change is not something that might happen in the future – it's going on right now,

How do we know that climate change is happening?

all around us. For millions of people suffering from droughts, floods and storms, its impacts are already very real. The Global Humanitarian Forum, an agency led by ex-UN Secretary-General Kofi Annan, carried out an assessment in 2009 into the human cost of climate change. It found that 300 million people were

---

Skeptics' Corner: Extreme weather

The Antarctic ice has been growing, not shrinking! This is true for certain bits of Antarctica – and, intriguingly, it actually gives us *more* evidence for global warming. Increased evaporation caused by higher temperatures has led to more snowfall in some parts of Antarctica, which has led to a thickening of the ice in these areas (usually inland). Elsewhere in Antarctica (mainly the bits near the sea, which is warming faster than the air) the ice is retreating. As we've noted before, it's a complex picture!

Climate change has good effects, not just bad ones! Again, this is true – up to a point. Warmer winters in some countries will mean fewer people die from cold. Some places are becoming easier for growing crops thanks to increased temperatures. Sadly, these small pockets of positive effects are hugely outweighed by the negative ones.[64] We have built almost all our settlements in places that are comfortable and fertile in our current climate, and our ways of growing food are carefully adapted to the weather we've been used to for thousands of years. Rapid changes in the Earth's climate are already starting to throw this out of kilter – hence the deaths from storms, floods and famines documented above. Even if climate change makes some places more comfortable to humans, do we expect everyone in the world to migrate to the places where climate change isn't so bad? What would happen if billions of people found that their countries had become inhospitable and that they needed to move to a new one? It's always worth thinking these things through...

Climate change is a problem, but there are bigger problems that we need to tackle first: This is an argument famously put forward by the Danish statistician Bjorn Lomborg. It ignores several important facts. First, it's a false choice – we can (and should) tackle such problems as climate change, poverty and health all at the same time. In fact, many of these problems have the same root causes (as we'll see later), so it makes sense to work on them all together. Second, climate change is making many of the world's problems much worse – it threatens all eight of the UN Millennium Development Goals. Any progress we make in these areas will be swept away by climate change, unless we act to prevent its worst effects. ■

already feeling the effects, through the loss of their homes, crops, lands or livelihoods; it also estimated that 300,000 people per year are currently being killed by climate change.[63] Of course, these are just estimates – it's hard to attach precise numbers to such a complex picture – but they do give us some idea of the scale of the crisis that's quietly unfolding all around us.

5) All these things are connected

So, to summarize what we've established so far:
• Carbon dioxide is a greenhouse gas. It has a warming effect on the planet.
• We've been pumping lots of extra carbon dioxide (plus some other greenhouse gases) into the atmosphere for the last 150 years. It's now at a level normally associated with much hotter prehistoric times.
• The air, land and oceans of the earth are heating up.
• We can measure this temperature change directly, and also see its effects in increased weather-related disasters, melting ice sheets and changes in wildlife behavior and numbers.

It's not hard to join the dots here, and conclude that humanity's greenhouse gases are causing the planet to heat up and the climate to change. But how can we be *totally* sure that these things are linked?

There are several very good reasons to make this connection:

* Why would all our CO_2 *not* be warming the planet?
Why on earth would this powerful greenhouse gas, that we know has a warming effect and has led to major climatic changes in the past, suddenly stop working? We've increased it by 40 per cent from pre-industrial levels, to a level associated with prehistoric sea levels many meters higher than today, and we're supposed to believe it's having no effect? It's a pretty incredible thing to claim and would need a pretty amazing

explanation, with evidence to back it up. No-one has yet presented any such explanation or evidence.

* There's nothing else that could be causing it

Similarly, if CO_2 isn't causing the warming, then what on Earth is? It's true that other things have warmed the planet in the past – kinks in the Earth's orbit, increased solar activity, changes in ocean circulation, volcanic activity. But there hasn't been enough of any of these things over the last 150 years to account for all of the warming we're seeing,[65] so those things can't explain the current bout of climate change.*

* CO₂'s fingerprints are all over the crime scene

We've already seen that temperatures and CO_2 levels have been rising together throughout the 20th century. That's decent circumstantial evidence that puts CO_2 in the right place at the right time, but what if we want something firmer? Well, we can also use what's known as the *thermal signature*. Using satellites, we can measure and analyze the wavelength of the Sun's rays as they fall onto, and are reflected back from, the Earth. There are specific wavelengths of radiation that are stopped by greenhouse gases like CO_2 and methane, and we'd expect to see fewer of them bouncing back up through the Earth's atmosphere if there really were an increased greenhouse effect. In other words, when CO_2 and methane trap heat in the atmosphere, they leave a recognizable 'signature' in the surplus rays that make it back out into space.

Recent, repeated satellite measurements have found this signature, growing at just the rate we'd expect.[66] That's solid evidence that increased levels of CO_2 and methane are trapping extra heat within the Earth's atmosphere.

---

* It's worth noting that these 'natural' effects are causing *some* warming, but aren't powerful enough to account for most of what we're seeing at the moment.

Skeptics' Corner: The link between CO_2 and global warming

But what about global warming in prehistoric times or on other planets? That can't have been caused by humanity's CO_2 emissions! Well, no, of course not – no-one's saying it was. Planetary temperatures have risen and fallen for all sorts of different reasons in the past (or on Mars or Venus) but here and now it's humanity's greenhouse gases that are heating up the Earth rather than orbital cycles or geothermal effects.

It's all about the Sun/sunspots/solar flares: The Sun does occasionally go through periods of increased activity, where it emits extra heat. However, all the extra sun activity of the last 150 years can only account for a small amount of the warming we've seen in that time – the rest must be due to something else. Since 2003, the Sun has, in fact, been cooler than usual. This is all well measured and non-controversial.[67] So climate change can't be due to the Sun alone.

It's cosmic rays/something else: There are other 'theories' out there, but with little or no evidence to back them up. Of course, in order to overturn the existing theory of human-made global warming (as detailed in this chapter), they'd need to have a huge mountain of contrary evidence that explains why the current scientific explanation isn't correct. For example, if you decided that the *real* reason why all the leaves fall off the trees in winter is that they're being tugged off by mischievous squirrels, you'd need a bit more evidence than a photo of a squirrel sheepishly clutching an oak leaf. You'd need a lot of new proof to back up your hypothesis* and some very good reasons to discount all of the evidence for the prevailing theory about why the leaves come off the trees, before making that sort of claim. None of the other supposed explanations for climate change can do this.

Very thin climate change 'explanations' sometimes get more prominence than they deserve, when publicized by a credulous journalist or unscrupulous TV show. Always ask: who is putting this alternative explanation forward, and what do they stand to gain? There are some excellent science websites that will usually jump on any new hypotheses quite quickly, explaining what they mean and whether they have any credibility: www.realclimate.org and www.skepticalscience.com are usually good places to start. ∎

---

* Strictly speaking, a 'theory' is something that has serious evidence backing it up and has become generally accepted as a good scientific explanation (e.g. the theory of evolution, the theory of human-induced climate change). By contrast, a suggested explanation that has yet to be rigorously tested is called a 'hypothesis'.

How do we know that climate change is happening?

We're left with the inescapable conclusion that humanity's greenhouse gases are causing the climate change we see around us. Which, if you think about it, is a bit of a relief – it means we can do something about it. If all this warming and weather chaos was being caused by forces outside our control, it would be a pretty bleak outlook for humanity. Luckily, it's being caused by greenhouse gas emissions – and we can do something about those.

1 J Fourier, *Mémoires de l'Académie Royale des Sciences*, 1827. There's a translation at http://nin.tl/bxSjtU **2** J Tyndall, *Philosophical Transactions of the Royal Society of London*, Vol 151, Part I, 1861. There's also a short biography of Tyndall at http://nin.tl/cUVbGu **3** S Arrhenius (1896). Available online at http://nin.tl/9td4cy **4** See a BBC video demonstration at http://nin.tl/cHrlqs **5** See http://nin.tl/bew6i5 **6** You can see the latest data from this station, and a full explanation of how it is collected and calculated, at http://nin.tl/b1qTU0 **7** IPCC Fourth Assessment Report (2007). **8** National Oceanic and Atmospheric Administration, http://nin.tl/a7Fsz0 **9** Petit et al, *Nature*, 399, 1999. http://nin.tl/amWq9Z **10** GS Dwyer, and MA Chandler, Phil Trans Royal Soc A, 367. **11** See footnote 7. **12** According to the International Energy Agency, global cement production produces 1.8 billion tonnes of CO_2 per year, of which half is from energy use and half from the chemical process involved. **13** World Resources Institute, http://nin.tl/ahQXSb **14** See, for example, 'Climate change skeptic Ian Plimer argues CO_2 is not causing global warming', *Daily Telegraph*, 12 Nov 2009. **15** http://nin.tl/9N4i5B **16** LeQuere et al, *Nature Geoscience*, Nov 2009, http://nin.tl/csqmyt **17** http://nin.tl/bk59Hd **18** http://nin.tl/d4vx46 **19** All of these processes are explained in more detail in James Hansen, *Storms of my Grandchildren*, Bloomsbury, 2009. **20** *The Guardian*, 12 Aug 2010. **21** Associated Press, 12 Aug 2010. **22** *The Australian*, 16 Jul 2010. **23** *The Washington Post*, 30 Aug 2010. **24** Data from NASA – see http://data.giss.nasa.gov/gistemp for the latest numbers. **25** As can be seen from the ice core graph on page 20. **26** See footnote 24. **27** NOAA State of the Climate Global Analysis, Feb 2007. **28** 'The races will go on (with imported snow)', *New York Times*, 19 Jan 2007. **29** 'Climate Change vs. Mother Nature', *The Independent*, 21 Dec 2006. **30** 'Spring flowers face extinction', *Daily Mail*, 18 Dec 2007. **31** National Snow and Ice Data Center, http://nsidc.org **32** Zwally et al, 'Surface Melt-Induced Acceleration of Greenland Ice-Sheet Flow', *Science*, 8 Apr 2002. **33** 'Ice sheet in Greenland melting at record rate', *Daily Telegraph*, 13 Aug 2010. **34** The United States National Oceanographic and Atmospheric Administration (NOAA), the United States National Aeronautics and Space Administration (NASA) and the Climate Research Unit (CRU) at the University of East Anglia in the UK. **35** Jerry Silver, *Global Warming and Climate Change Demystified*, McGraw Hill, 2008, contains a useful short history of temperature measurement. **36** http://www.surfacestations.org/ **37** http://nin.tl/btoYzr **38** See, for example, http://nin.tl/b9wELP **39** See, for example, G Stanhill and S Cohen, 'Global dimming', *Agricultural and Forest Meteorology*, 107, 2001. **40** House of Commons Science and Technology Committee (31 Mar 2010), Science Assessment Panel (14 Apr 2010), & Independent Climate Change Email Review (7 July 2010). **41** Met Office, 2009, http://nin.tl/dyuKWv **42** George Marshall, *Carbon Detox*, Octopus, 2007.

43 For more detailed explanations of how these different effects are driven by climate change, see the FAQ section of the IPCC's Fourth Assessment Report at www.ipcc.unibe.ch **44** Münchener Rückversicherungs-Gesellschaft. To access this data, register at https://www.munichre.com **45** http://nin.tl/ci2SjX **46** 'Global warming blamed', *The Australian*, 12 Aug 2010. **47** WMO, 11 Aug 2010, http://nin.tl/d5ZqrS **48** Vital sources for this section included: Kirstin Dow and Thomas E Downing, *The Atlas of Climate Change*, Earthscan, 2006; Jerry Silver (see footnote 35); George Marshall (see footnote 42; 'Climate Justice', *New Internationalist*, 419, Jan/Feb 2009. **49** The Pew Climate Center explains this well at www.pewclimate.org/hurricanes.cfm **50** Zhang et al, 'Detection of human influence on 20th-century precipitation trends', *Nature*, 448, 2007. See also the IPCC Fourth Assessment Report (2007). **51** See footnote 7. **52** Arctic issue, *New Internationalist*, 424, Jul/Aug 2009, www.newint.org **53** Konrad Steffen, CIRES, University of Colorado at Boulder. **54** 'Greenland ice sheet faces tipping point in 10 years', *The Guardian*, 10 Aug 2010. **55** 'Sea level rise could bust IPCC estimate', *New Scientist*, 10 Mar 2009. **56** 'The Future is Here: climate change in the Pacific', Oxfam Australia report 2009. **57** Schliebe et al (2006), http://nin.tl/aqP6D7 **58** E Kolbert, *Field Notes from a Catastrophe*, Bloomsbury. 2006. **59** This is widely documented around the world. For a recent scientific assessment of the phenomenon see Pulido & Berthold, *Proceedings of the National Academy of Science*, 107, 16, 2010. **60** Zoological Society of London 16 Jul 2009. **61** Pounds et al, 'Widespread amphibian extinctions from epidemic disease driven by global warming', *Nature* 439, 2006. **62** See footnote 48. **63** Global Humanitarian Forum, 2009, http://nin.tl/awh8nC **64** There's an excellent list of the positive and negative effects of climate change, with links to the relevant scientific research, at http://nin.tl/bBzYPb **65** See footnote 7. **66** Harries et al, *Nature* 410, 2007; Griggs and Harries, *Proceedings of SPIE: Infrared Spaceborne Remote Sensing XII*, 5543, 2004; Chen et al, paper presented to EUMETSAT conference 2007, http://nin.tl/bSDFP1 **67** See footnote 11.

2 How bad could it get?

What the world will look like if temperatures continue to rise... The feedback effects that could take warming out of our control... And why there's no time to lose.

SO WE'VE LOOKED at how we know climate change is happening, its causes, and the effect it's had so far. We now need to ask ourselves: what about the future? If we do nothing about this problem, how bad could it get?

To assess this, climate scientists have been building increasingly complex models, to map out different possible future scenarios and give some indication of what we can expect. This is one of the trickiest areas of climate science – the climate system is incredibly complicated, and everything depends on how much greenhouse gas our society emits, and how fast we emit it. As we'll see, there are also all sorts of tipping points and feedbacks that might come into play.

On one level, we don't need these models at all. We've already seen the effects of raising the Earth's temperature by 0.8°C, and they're not pretty. The more greenhouse gas we pump into the air, the worse the warming will get – and so will all the other knock-on effects, such as floods, storms and droughts. Basic common sense tells us that we need to reduce our greenhouse gas emissions as fast as we possibly can. We don't need a computer model to understand this basic point.

We are also being told a very powerful story by the prehistoric evidence – that rising CO_2 levels can trigger feedback loops, leading to huge changes in global temperatures and sea levels. There's no reason to believe that this won't happen again if we carry on the way we're going. Again, we don't need to use computer models to realize this.

But how much do we need to reduce our emissions in

order to avoid the worst climate effects? How long have we got, and what exactly will happen if we don't act? To answer these questions, we need a bit of scientific prediction, which is where the IPCC comes in.

The Intergovernmental Panel on Climate Change (IPCC) was established by the UN in 1988, to assess 'the scientific, technical and socioeconomic information relevant for the understanding of the risk of human-induced climate change.'[1] It went on to issue a series of major climate science reports.* Its predictions about future impacts represent an extraordinary consensus from around 2,500 climate scientists – the IPCC can be said to represent the scientific middle ground. Their predictions are, therefore, fairly cautious and conservative. Former chair of the IPCC scientific assessment panel, Sir John Houghton, said that they had 'deliberately underestimated the problem'. Despite this, the report still makes pretty grim reading, as you can see from the highlights (lowlights?) below.

Greenhouse gas emissions
The IPCC's projections are based on a range of different greenhouse gas 'scenarios'. There's a low-end scenario, where global greenhouse gas emissions rise gently until 2040, then trail off and start to fall. There's a high-end scenario, where emissions keep rising rapidly and peak in 2080. There are a few others in between. These scenarios all assume that no policies are put in place to reduce global emissions – they're an attempt by the IPCC to plot what the future would look like if we carried on with 'business as usual'.

Since these scenarios were first drawn up around the year 2000, humanity's emissions have been rocketing up in line with the 'high end' scenario. We're currently heading steadily along the IPCC's

---

* In 1990, 1995, 2001 and 2007.

How bad could it get?

worst-case path. I've therefore pulled out the predictions from the latest (2007) IPCC report that reflect this. The following describes a few of the things that the world's scientists think are *likely to happen* in the next 90 years if our emissions keep climbing the way they are now.

Global temperatures

With our current emissions path, there's an 80-per-cent chance of a global average temperature rise of between 2.4°C and 6.4°C by 2100, with a 'best estimate' of 4°C. Two degrees would take us past several dangerous 'tipping points' (see below), four degrees would eventually leave us with a radically different planet. Six degrees could lead to the extinction of most life on Earth. The remainder of this section presents the risks associated with a 4°C rise. This is five times the human-caused warming we've had so far.

Raising the temperature in this way pumps more and more energy into the climate system – and it's all got to go somewhere. That doesn't just mean temperature rises, it means a greater risk of fiercer storms and floods, and more evaporation leading to droughts and rainstorms. I sometimes think of all the different weather-related disasters (floods, storms, landslides, droughts, heatwaves) as being a bit like a group of crazed monkeys who've run amok in our house. They run around causing chaos and destroying things, we can't quite predict exactly what they'll do, but we *do* know that they get faster and wilder with every banana we feed them. The first sensible step is probably to stop pouring truckloads of bananas in through the windows.

While the following predictions can seem scary, they also *make sense*. These are the kind of things we'd *expect* to happen if we pushed lots of extra energy into the climate system – many of them are upsetting, but they shouldn't be too surprising.

Effects on the poles

Arctic sea ice is expected to vanish completely every summer, with serious knock-on effects for wildlife and indigenous peoples' traditional lifestyles. The Greenland and West Antarctic ice sheets are likely to undergo partial melting, raising sea levels between four and six meters. The precise timing of this is hotly debated – some scientists say it could happen this century, while others put it hundreds of years away – but, so far, the ice in the Arctic and in Greenland has been melting faster than predicted by the IPCC.

Effects on low-lying coasts and islands

Around 10 per cent of the world's population – almost 700 million people – live less than 10 meters above sea level. Any rise in sea level increases the frequency of storm surges and periodic flooding, putting people's homes, crops and livelihoods at risk.

So far, the IPCC has only predicted a slow rise in sea level this century – a continuation of the current rate of three millimeters per year. However, the unexpectedly rapid melting of polar ice since 2007 has caused a number of scientists to challenge this forecast. The latest figures suggest a rise between 0.75 and 1.9 meters this century if we carry on down our current emissions path.[2] Just a one-meter rise in sea level would permanently flood around 60 million people out of their homes, mostly in poorer countries. Much of Bangladesh, Vietnam and the Nile delta would be underwater. The Netherlands – which is already struggling with sea level rises of three millimeters per year – would be forced to abandon much of its country to the sea, and we could probably say goodbye to Venice. Low-lying islands such as those in Tuvalu, Kiribati and the Maldives would probably need to be completely evacuated.

How bad could it get?

Effects on drinking water
One sixth of the world's population relies on water supplies fed by mountain glaciers. In a 4°C warming scenario, many of these glaciers are expected to shrink rapidly, leading to severe water shortages for millions of people and vast areas of farmland. Regions at particular risk include China, Peru, Pakistan and California.

Effects on weather
At 4°C of warming, we could expect once-in-a-lifetime disasters to crop up much more frequently. In Europe, the infamous 2003 heatwave would become a normal summer event. Combined with reductions in freshwater, these recurrent disasters would almost certainly lead to a decline in global food production, resulting in mass famine on an unprecedented scale.

Effects on health
The IPCC predicts that a 4°C rise would lead to many more deaths from malnutrition, drought, excessive heat, drowning and disease. Infectious illnesses such as malaria and dengue fever would creep steadily northwards as the creatures that carry them move with the changing climate.

Effects on wildlife
At least 40 per cent of the world's species would be doomed to extinction in a 4°C scenario. Most coral reefs would be killed, and there would be a serious risk that the Amazon rainforest would dry up and start to burn down from the year 2040.

Extra post-2007 effects: Over-tipping
Most of the predictions above come from the IPCC's last report, which was in 2007. Not only were its predictions generally agreed to be on the conservative side, it's also rapidly going out of date. Their next report is due in 2014; in the meantime there are two

key areas that their 2007 report only touched on, but are considered by many climate scientists to need urgent attention:

First, the ice in the Arctic Ocean and on Greenland is melting faster than the IPCC's predictions. Scientists are still debating the reasons for this – it seems that there may be extra feedback loops at play in addition to the albedo effect mentioned in Chapter 1.* This speedier thawing process matches neatly with the prehistoric data – as temperatures rose at the end of ice ages, the ice at the poles melted surprisingly fast, suggesting the existence of some powerful feedback mechanisms that we haven't quite worked out yet.[3] This in turn suggests a much higher risk of losing large pieces of the Greenland and Antarctic ice sheets this century.

The result of that would be a sea level rise of up to six meters in our lifetimes, or the lifetimes of our children. This would put large parts of many major cities underwater, including New York, London, Sydney, Vancouver, Mumbai and Tokyo. The effects on coastal populations around the world would be catastrophic.

(Not very) positive feedbacks
Second, the IPCC's 2007 report only briefly mentions the risk of *runaway climate change* caused by *positive feedbacks*. A positive feedback occurs when something caused by global warming has a knock-on effect that further speeds up the warming process. Here are some of the most important feedback effects that might be facing us this century:
• **Melting ice.** We've mentioned this one already – as snow and ice melts in Greenland and the Arctic Ocean, the white reflective surface gives way to reveal the much darker land or water underneath. This

---

* For example, as the great glaciers of Greenland start to thaw, the meltwater they produce seems to be speeding up the melt by cutting runnels in their surfaces, and also creating lubricated 'beds' underneath the glaciers that may make it easier for pieces to slide off into the ocean.

absorbs more heat from the sun, thus speeding up the warming of the planet.

• **The release of methane from the permafrost.** Billions of tonnes of methane, a powerful greenhouse gas, are stored in the frozen tundra of Siberia, Canada and Alaska. As these areas warm, they start to release the gas, thus adding to the total global warming effect. Methane emissions from these areas have risen by almost a third in the last five years.[4] There are also huge amounts of stored methane in the cold depths of the ocean, which are likely to be disturbed if we allow the seas to keep on heating.

• **Vanishing forests.** If the world warms by more than 2°C, there is a serious risk that the rainforests will dry up and begin to burn. This will release vast amounts of CO_2 into the atmosphere, hastening warming even further.

• **Less ocean absorption.** Stormier weather, increased temperatures and greater acidity (from so much dissolved CO_2) could have an effect on the ocean's ability to suck up carbon dioxide. These relationships are complex, however, and still not well understood.

• **Water vapor.** Warmer temperatures increase evaporation, putting more water vapor into the air. Water vapor is a greenhouse gas, so this process adds to global warming.

If we don't curb our emissions, the effects of these feedbacks will probably become so strong that they will keep on feeding themselves and adding to global warming no matter what we do. This is what happened in those prehistoric periods of rapid warming – the initial temperature increase created feedback effects, which eventually reached a 'tipping point' – a point of no return – and global warming 'ran away with itself', way beyond the level that would have been caused by the initial warming event alone. If we don't curb our emissions urgently we'll start hitting our own climate tipping points, triggering similar unstoppable runaway change.

To avoid riding this runaway climate train to a *really* scary future, it looks like we'll have to cut our carbon emissions further and faster than the IPCC suggests (they openly admit in their 2007 report that they have not taken all of these feedbacks into account in their projections). There's more about carbon-cutting targets in Chapter 5.

So to summarize: if we don't take action to prevent climate change, there is a *high likelihood* of some really unpleasant and tragic scenarios, and the *serious possibility* of some hugely catastrophic events. Again, it's all a bit much to take in; it all seems too awful to be true. I hate to tell you this, but I've purposely been a bit cautious with the descriptions above, and tried to leave out some of the more extreme examples of possible climate change outcomes. All of the above is based on solid, well-documented scientific research. Remember: although we're already committed to *some* warming from all the greenhouses gases that have already gone into the atmosphere, there is still time to avoid the worst of the effects listed above. Remember too, that most of the solutions would improve our lives in lots of other ways as well.

If you want to read a really scary account of how bad it might get, pick up a copy of Mark Lynas's 2007 book *Six Degrees*. Make sure you're not reading it alone in the dark, though.

Uncertainties

While the 'big picture' climate change predictions (increased temperatures, more evaporation, melting ice caps and rising seas) are based on observed results and the prehistoric record, the more detailed projections (how much climate change, where, and by when) are based on computer models. These models are constantly checked and improved, and tested against real-life scenarios to make sure they're as accurate as possible, but there are always going to be some uncertainties. This is why climate scientists talk in terms of probabilities and risks; no-one knows all the details of what's going to happen as the climate changes, but these models can show us the most likely trends and give us some useful indications and warnings.

Some commentators say that, because of this

Skeptics' Corner: Future scenarios

Some opponents of climate action argue that things aren't going to be as bad as the scientists think, because there are negative as well as positive feedbacks – in other words, as temperature and CO_2 levels rise, there are processes that will kick in to slow down the warming. As is so often the case, there is some truth in this – but none of these feedbacks have been powerful enough to change the overall warming trend.

For example, some skeptics argue that CO_2 helps plants grow faster, and so all our emissions will act as a fertilizer, and all those extra plants will soak up the excess CO_2. This is a lovely idea, but not borne out by the facts: there's no sign of this 'fertilizer feedback' actually taking place on any significant scale. This is probably because CO_2 only speeds up plant growth if the plant also has everything else it needs to grow bigger – water, soil nutrients, light and space. This usually isn't the case. Meanwhile, climate change is already restricting the amount of water available to many plants.

There are *some* unknown factors which could have an effect on the overall warming trend – for example, clouds help to cool the planet down, and we still don't fully understand how they will react to a changing climate. However, there are other uncertainties that work in the opposite direction – for example, we've been emitting tiny particles called aerosols from factories and power stations, which have a cooling effect on the planet. These are gradually being reduced as industry standards are tightened, but because we don't know exactly how much we've emitted so far, we don't know how much they've been slowing down the overall global warming trend. If they've been having a big cooling effect, then as aerosol emissions fall global warming will speed up significantly.

None of these uncertainties (on either side of the equation) are likely to make a major difference to the overall warming trend – at best, they might buy us some extra time – and are we really going to gamble everything on the hope that a load of extra clouds might suddenly swoop in and save us? We can see the climate crisis unfolding around us, and we'd be extremely foolish not to act to prevent it – especially when so many of the solutions will help us in other ways too. ■

uncertainty, we shouldn't take action to tackle climate change. Why spend money on climate solutions, they argue, when it might not be as bad as we think? The problem with this argument is that we *do* know it's going to be bad. We know this because it's *already* bad – people are suffering in floods and droughts, we're losing species left, right and centre, and 300,000 people are dying every year. We don't need the models to tell us that if we keep on pumping out the polluting gases that caused this mess, things are going to keep getting worse – especially if we're churning them out faster than ever.

Just because we don't know the precise details of the problems we're going to face, doesn't mean we should ignore them. It's like saying 'it's fine to release these 100 wild rats into my kitchen – I don't know exactly what problems they're going to cause, and so there's too much uncertainty for me to waste money on taking them somewhere else'. As with a kitchen full of rats, we don't know exactly what's going to happen but we're likely to have a series of disastrous outcomes (all the food and kitchen equipment ruined), and there's the chance of a major catastrophe (a chewed wire burns the house down, or your whole family contracts Weil's disease). The sensible thing to do is to stop releasing rats into your kitchen.

1 www.ipcc.ch 2 Vermeer & Rahmstorf (2009), www.pnas.org 3 James Hansen, *Storms of my Grandchildren*, Bloomsbury, 2009. 4 Bloom et al, 'Large-Scale Controls of Methanogenesis Inferred from Methane and Gravity Spaceborne Data', *Science*, 327. no. 5963, 2010.

3 Why is climate denial on the rise?

The difference between skepticism and denial... Why it's easier not to believe... Misleading the public for fun and profit... How to spot a climate denier... The nonsense guide to climate change.

IF YOU'RE A SKEPTIC, then I salute you.

Skeptics are people who don't take things at face value; they demand facts to back things up, and are ready to change their opinions based on the weight of evidence, even if that goes against their personal preferences or beliefs. All good scientists are skeptics, and there's plenty of healthy skepticism within serious climate research – this is an important part of how science works, with scientists trying to challenge their own and each others' hypotheses by gathering relevant data. I like to think that I'm a bit of skeptic, too (although I'll need a bit more evidence before I'm sure).

Deniers, on the other hand, are people who refuse to accept evidence that conflicts with their personal beliefs, desires or ideology. People in denial gather reasons and excuses, however flimsy they may be, that allow them not to believe in whatever unwelcome truth they're trying to avoid. This is something that we're all guilty of from time to time – pretending that a piece of bad news isn't really that bad, taking time to come to terms with difficult problems or personal tragedies.

No serious skeptic in possession of all the facts could doubt that human-caused climate change is real, and dangerous – as we saw in Chapter 1, the evidence is just too overwhelming. However, there are many people who are still in a state of denial over climate change, for a wide range of reasons – in fact, I'd argue that pretty much all of us are at least slightly in denial over how serious this problem is.

Say it ain't so

It's easy to understand why so many people don't want to believe in climate change. Even those of us who accept, on a factual level, that it's happening and that it's serious don't seem to be changing our behavior in response to the crisis. Even though climate doubt has grown a little, 91 per cent of the British population still think that climate change is happening to some extent, and at least 60 per cent think it's serious.[1] An August 2010 poll in the US found that 60 per cent of people wanted the government to crack down on greenhouse gases from power stations and refineries.[2] Despite this, most of us still drive cars, use more energy than we need, and fly off on holiday whenever we get the chance. Marches and demonstrations demanding climate action from the government have never been bigger than tens of thousands of people. This pattern holds true for most of the industrialized world. It seems as though most of the people who claim to care about climate change still aren't doing very much about it. Why is this?

It's probably partly to do with the fact that low-carbon alternatives such as good public transport or well-insulated homes still aren't easily available (see Chapter 6), but it's almost certainly also because of denial. We tell each other that climate change is a serious issue, but we still don't really *believe* it. The British climate campaigner George Marshall has written a lot about this topic – in his book *Carbon Detox* he notes a number of reasons why we find it so hard fully to accept the reality of climate change:

• We've evolved to respond to dangers that are 'visible, immediate, comprehensible, involve personal physical damage and have a clear cause'. Climate change has none of these qualities. It's not a clear, present or obvious threat. It's invisible, gradual, and its effects are hard to predict. It just doesn't properly push our danger buttons.

Why is climate denial on the rise?

• There's no single, obvious enemy to blame it on. In fact, the uncomfortable truth is that we, the people of the industrialized North, are all culpable to some extent. That's very difficult to swallow – the idea that simple, innocent things we do every day could be causing harm to other people. Who'd want to believe that?

• A lot of the things we're told about climate change relate to the future – 20, 50, 100 years from now. This makes it seem even more distant and unconnected to us.

• We're bombarded with contradictory messages. The mainstream media intersperse reports of upcoming climate doom with adverts for cheap flights and new cars. Governments and campaign groups tell us that it's huge and serious, but that we can solve it by doing 'just a few easy things' like pumping up our car tires and reusing plastic bags. None of this makes much sense.

• Climate change has become associated with 'environmental' and 'green' issues. Many people don't think of themselves in these terms, and feel that environmental issues – though important – are 'one of those things that other people are sorting out'. In reality, climate change is not about 'the environment' – it's about people, our homes and livelihoods, and will fundamentally affect all of us at some point.

• It feels like a direct challenge to our familiar lives and lifestyles. So many of the messages about climate change are to do with giving things up or using less, and it's natural to want to kick back against this. It's true that our overall consumption of energy and materials will need to fall (see Chapter 6), but that's not the full story – the best and most effective solutions to climate change should have positive impacts on our lives and on society as a whole. Sadly, that's not a message we tend to hear very often.[3]

So when someone pops up on the TV or the radio with some plausible-sounding reasons why climate change isn't really happening, or isn't that bad, it's hard not to fall for it. Most of us *want* to believe that climate change isn't real, isn't serious, or has nothing to do with us.

Of course, this only applies to those of us living relatively comfortable lives in industrialized nations. The millions of people on the climate frontline, watching their homelands flooding, melting, drying out or being ripped apart by storms, don't have the luxury of denial.

If we're going to deal with the denial problem, campaigners and scientists need to be much better at communicating the climate reality. We'll come back to this a bit later in the book. Meanwhile, what about those professional, well-spoken, plausible people who keep appearing in the media to denounce the climate science? Who are they, and where do they fit in?

Strange growths in the hothouse

The public debate over climate science has become very weird lately.

The debate amongst genuine climate experts (people who've worked on climate science all their lives and published professional papers on the subject) is all about the details: How much human-induced warming is going to happen, by when? What will the precise effects be, and what do we need to do to prevent it? The basic science linking humanity's greenhouse gas emissions with climate change is as well-established as the link between smoking and lung cancer, or between HIV and AIDS – some people still deny these connections, but no-one takes them seriously. According to a 2009 survey, 97 per cent of published climate scientists believe that humanity is changing the climate.[4]

You wouldn't know this from watching or reading the mainstream media. Newspapers still routinely refer to climate science as 'controversial'; TV and radio debates frequently feature commentators questioning the basic principles of climate change. I'd be very surprised if you hadn't previously encountered at least some – and maybe all – of the issues raised in the 'Skeptics' Corner' boxes in the previous two chapters.

They're often being put forward by the same, relatively small group of prominent self-styled 'climate skeptics' – journalists, authors, politicians, and even one or two scientists who seem to have devoted their lives to challenging climate science. It's not hard to imagine why they do this. Most of them probably believe what they're saying and are in denial about climate change, as so many of us are.

Some also seem to be doing it for ideological reasons, because preventing total climate chaos will require some major changes in society. We'll need to stop burning fossil fuels, rein in the more destructive activities of corporations, governments and markets, give more power to disenfranchised communities and make some fundamental changes to the way our economies work (see Chapters 8 and 9). This strongly conflicts with the political ideology, business interests, and/or funding sources of certain politicians and commentators. Rather than changing their political beliefs to fit the facts, they've chosen to deny the facts instead.

Other professional deniers may simply be coming from the same place so many of us are – climate change is scary and threatens to interfere with their comfortable lives and so they don't want to have to accept it. You'll note that these public contrarians are usually white, wealthy, middle-aged or older men. They've generally done well out of the status quo and are about as far away from the climate frontline as it's possible to get.

Whatever the reasons, the tactics are usually the same – to find some plausible-sounding reasons why climate change isn't happening, or isn't serious, or isn't caused by humans, and then repeat them again and again at every opportunity. Some public deniers (like Christopher Monkton[5] and Ian Plimer[6]) use made-up facts and figures to confuse the debate; others (like Richard Lindzen[7] or Christopher Booker[8]) use more sophisticated tactics, focusing in on small uncertainties in climate science, or minor errors and improprieties made by climate scientists or campaigners, and making them seem more significant than they really are. Very few of their claims ever stand up to serious scrutiny, but that doesn't stop them from repeating them again and again.[9]

It's no surprise that these people exist – if you look hard enough, you can find people willing to denounce anything in public. There are people who think the British royal family are really a group of seven-foot lizards in disguise.[10] The real question is – why are the climate deniers so prominent? Why do they get so much airtime, sell so many books, address major conferences and even get voted into public office? If a newspaper columnist started writing weekly articles saying that all the doctors were wrong and that smoking was really good for you, they wouldn't be taken seriously – so why do we put up with journalists writing equally deadly nonsense about climate change?

Misleading the public for fun and profit

All the public deniers have one important thing in common: none of them believe that we need to reduce CO_2 emissions. This makes them incredibly useful to anyone who wants to keep emitting CO_2 – in particular, the big energy corporations, whose short-term profits rely on burning lots of fossil fuels. Look behind many prominent climate deniers and you'll find a web of fossil fuel funding. For example, ExxonMobil pours millions of dollars into anti-climate-science 'think tanks' and lobby groups.[11] The Institute of Public Affairs, an Australian group that frequently sponsors climate deniers, receives large amounts of funding from coal, oil and gas companies.[12] All the evidence suggests this is a deliberate tactic by the energy corporations – they don't need to come up with any serious scientific arguments against human-made climate change, they just need to spread enough confusion and misinformation to slow down action on climate change, thus maintaining their profits for a little while longer. It's the same tactic that tobacco companies used to delay government action on smoking for so many years (and is even being carried out by some of the same people[13]).

Right-wing organizations and wealthy individuals are also pouring money into climate denial – they don't like the fact that tackling climate change means taking power away from corporations and giving it back to the public. For example, the Heartland Institute, an American think-tank that sponsors and promotes climate deniers, receives large amounts of funding from private trusts that seek to deregulate business and privatize government services.[14]

Meanwhile, telling the public what they want to hear – that climate change isn't that serious – is a great way to sell lots of books or newspapers, or just to gain notoriety. Books by climate contrarians such as Bjorn Lomborg and Fred Singer sell depressingly well, denialist articles generate measurable jumps in the readership of online newspapers, and Christopher Monkton's widely debunked anti-climate-science lecture has millions of views on YouTube.[15]

All of this means that there's a very good living to be made in climate-change denial. For example, Lord Monkton's two-week Australian tour in 2009 reportedly netted him around US$20,000.[16] Ian Plimer earned almost $400,000 in two years from sitting on the boards of mining companies.[17]

This isn't to say that the anti-climate-science brigade are simply in it for the money. I'm sure most of them believe in the things they say. The point is that they wouldn't have such prominence and status without the backing of certain wealthy and powerful individuals, political groups, media outlets and corporations with a (horribly short-term and profit-driven) interest in preventing action on climate change. They're also helped out by the general state of public denial referred to above. All of this is presumably self-reinforcing; the more publicity, funding and plaudits the deniers get, the more entrenched they must become in their unfounded beliefs.

The mainstream media also have a tendency to set up head-to-head 'debates' between climate scientists/campaigners and climate change deniers – not only is this supposed to be more exciting than getting into the real details of the science, it also gives the illusion of 'balance'. Of course, in reality, 97 per cent of climate scientists assert that climate change is serious and mostly caused by humans; but if all people see is one-on-one debates with climate change deniers, it makes it look as though the science is still disputed. It's about as useful as watching a debate on how to solve the African AIDS crisis between an experienced Ugandan health campaigner and someone who believes that AIDS is spread by evil pixies and can be cured by eating spaghetti – it doesn't take us anywhere helpful. It also means that we've been missing out on having the far more important debates – what are the best solutions to climate change, and how can we make them happen?

Science friction
But what makes climate scientists any better? Surely they have vested interests in all this too? Can we really trust them to tell us the truth about climate change?

In order to get anything published in a scientific journal, scientists need to back up their work with evidence – enough evidence to convince a separate group of scientists that their work has merit. This is the 'peer review' process, and it's a powerful tool for ensuring that published science is backed up by decent, repeatable research. This means that – unlike journalists, commentators, advertisers, campaigners, or politicians – scientists need to be able to support their statements with robust facts and data in order to advance in their profession.

The fundamental science of climate change has been built on a mountain of such evidence that's been put together over decades. Of course different climate scientists have different opinions on the details of the climate problem; competition between all these different ideas helps to keep scientists on their toes and to keep science moving forwards. Whoever can present the most convincing evidence is most likely to get their ideas accepted. So accepting the reality of climate change isn't about 'believing' the individual scientists – it's about following the *evidence*.

Amidst this lively and ongoing scientific debate about the details (though not the fundamentals) of climate change, the Intergovernmental Panel on Climate Change (IPCC) has the unenviable task of finding the common ground. The IPCC doesn't do any research of its own; instead, it collates all the relevant peer-reviewed research carried out by scientists all over the world.

The IPCC Fourth Assessment Report included more than 2,500 scientific expert reviewers, more than 800 contributing authors, and more than 450 lead authors. For example, the Working Group 1 report about the physical climate science (including the summary for policy makers) included contributions by 600 authors from 40 countries, and over 620 expert reviewers, a large number of government reviewers, and representatives from 113 governments.[18]

Some commentators accuse the IPCC of somehow conspiring to exaggerate the risks of climate change.

Exactly how such an extraordinary international conspiracy, involving thousands of scientists from 194 countries, could possibly be organized is never properly explained. In fact, because the IPCC focuses on the agreed common ground between researchers and skims over newer areas of research (such as the risk of positive feedbacks, more-rapid-than-expected ice melt and runaway climate change) there is a real chance that the IPCC could be *under*estimating the threat we face.

Oiling the wheels

A common accusation from anti-climate-action commentators is that scientists are exaggerating the threat of climate change in order to 'get more funding'. Think about this for a moment and you'll realize it's completely back to front – if a scientist found real evidence that contradicted the accepted theory of human-made climate change, do you think they'd have any difficulty finding funding? We've already seen the kind of money that commentators like Monkton and Plimer are being paid just to present their opinions, without the backing of any reputable science at all.

Money certainly can affect the direction of scientific research.* But most of the political and financial pressure on climate researchers has been pushing in exactly the opposite direction from the one that anti-climate-action critics suggest. A survey of climate scientists by the Union of Concerned Scientists in 2007 found that 58 per cent of respondents had experienced political pressure to water down their scientific findings.[19] The 2007 IPCC report was stripped of many 'undesirable' passages by politicians before it could be published, including warnings about the likely impacts of climate change on North America and references to

---

* For example, consider how much is spent by pharmaceutical companies on seeking cures for the minor health problems of the wealthy (baldness, wrinkles) as opposed to the major diseases of the poor (malaria, cholera).

positive feedback loops and the risk of runaway climate change.[20] The George W Bush administration in the US between 2000 and 2008 was particularly notorious – they gutted scientific reports, removed references to human-made warming and threatened the careers of government scientists who dared to speak out.[21] More recently, the outspoken IPCC chair Rajendra Pachauri has endured a barrage of false claims of fraud and corruption from climate deniers.[22] There is plenty of pressure on climate scientists to change their research – but nearly all of it is pushing them to tone down their message, and not to speak out.

How to spot a climate change denier

Here are some tips for distinguishing a climate change denier from a genuine skeptical scientist:

• They repeat arguments that have already been shown to be false. If they use any of the arguments from the 'Skeptics' Corner' boxes in Chapters 1 and 2, for example, that should tip you off.

• They attempt to pick holes in climate science without presenting any real evidence for alternative theories on what is happening to the climate.

• They focus on small mistakes made by individual scientists or campaigners, or institutions like the IPCC, and try to suggest that this discredits climate science as a whole.

• Their ideas often require some sort of huge, unlikely conspiracy in order to be true.

• They very rarely admit that they've ever been wrong. If one of their arguments becomes completely discredited, they will usually simply switch to another and never mention their earlier position again.

• They speak in snappy little soundbites. A genuine scientist is far more likely to say something long-winded and uninspiring like 'of course, there are many uncertainties but the balance of evidence strongly suggests a warming trend with a strong anthropogenic component'. Which translates as: 'yes, it's getting hotter, and most of it's down to us.'

A final, key point to make about our globe-trotting science-denying friends is the sheer incoherence and contradictory nature of their arguments. The IPCC is able to pull together a summary of all the current climate research because even though those thousands of scientists are all working in different areas of climate science, their work all adds up into a single coherent picture. The different bits of evidence all fit together, and support the central theory of human-induced climate change.

To illustrate how this differs from the anti-climate-science position, I've attempted to write a short synthesis of the key arguments from the most prominent climate change deniers, and show how they all fit together (or rather how they don't). I call it the Nonsense Guide to Climate Change (see overleaf).

1 Poll by Ipsos Mori in Feb 2010. **2** http://nin.tl/9sgY8j **3** George Marshall, *Carbon Detox*, Octopus, 2007. **4** PT Doran & M Kendall Zimmerman, 'Examining the Scientific Consensus on Climate Change', *EOS, Transactions of the American Geophysical Union*, 90, 2009. **5** J Abraham (2010), http://nin.tl/973KPn **6** I Enting (2009), http://nin.tl/bgyETo **7** See, for example, http://nin.tl/baK4zo **8** Robin McKie, 'What planet are they on?', *The Observer*, 9 Dec 2007. **9** There are numerous examples at www.desmogblog.com **10** Visit www.davidicke.com, if you dare... **11** Union of Concerned Scientists (2007), http://nin.tl/cgrktw **12** www.sourcewatch.org **13** See footnote 11. **14** See footnote 12. **15** See footnote 5. **16** 'Global warming skeptic takes his message to Noosa', *The Noosa Journal*, 11 Jan 2010. **17** www.prwatch.org/node/8686 **18** www.ipcc.ch **19** Union of Concerned Scientists and Government Accountability Project, Feb 2007, http://nin.tl/9dchGb **20** Roger Harrabin, *Today*, BBC Radio 4, 6 Apr 2007; David Wasdell, Feb 2007, http://nin.tl/bzS87k **21** James Hoggan & Richard Littlemore, *Climate Cover-Up: The Crusade to Deny Global Warming*, Greystone, 2009. **22** John Vidal, 'If Rajendra Pachauri goes, who on Earth would want to be IPCC chair?' *The Guardian*, 3 Sep 2010.

Why is climate denial on the rise?

The nonsense guide to climate change

Don't worry. The planet isn't warming up at all – the temperature record has been tampered with by dodgy climate scientists, in order to justify their funding, and anyway all the measuring devices are next to hot air vents in town centers. Actually, the planet has been warming, but it stopped in 1998 – that's why the sea ice has been growing this year. In the winter.

Having said all that, the planet is in fact warming rapidly, but it's nothing to do with CO_2 – it's all about sunspots and cosmic rays. Those crazy climate scientists were telling us there was going to be an ice age, and now they say we're warming up! Which of course we are. Except there was a record cold winter in my town this year – where's your global warming now, eh?

The fact that CO_2 is warming the planet is not in dispute – but humans make only a tiny contribution to it. The rest comes from volcanoes, and anyway if you look at the prehistoric record the warming came first, then the CO_2! Carbon dioxide is good for us, it makes plants grow. Plus it's warming on Mars, and it was warm in medieval times, I expect our 4x4s caused that too, did they?

The endless flaws and errors in the IPCC reports, plus the scandalous data from the hacked CRU emails, show that global warming alarmists are being used as the tool of governments to raise all our taxes and usher in a new communist world order. It's a huge hoax, and they're all in on it – the 7,000 measurement stations, the migrating birds, whoever's making Greenland melt, all those carefully staged natural disasters – they do it in the same studio where they faked the moon landings, did you know that? Plus, global warming will be great! We'll grow grapes in London and plant crops in the Arctic and no-one will freeze in the winter. Except it's not happening at all, and is all part of a natural cycle – the world has always changed temperature, you know. And it's all China's fault.

We can all adapt to it easily enough, and should be spending our time and money on the real problems. It's all based on computer models, for goodness' sake! Meanwhile, it's too late to do anything and we're all doomed anyway – we may as well party while we can. Because it's not happening, except it is, and it isn't our fault, although it is, but it's not that serious, except it is, but there's nothing we can do.

In summary: Relax. It's all going to be fine. You don't have to do anything. Don't look at the nasty scientific reports, or the people dying in floods and droughts and storms, or the melting icecaps, or the way the seasons have started getting really weird, or the strange insects showing up in your local park. Shhhhh. Settle down now. Everything's going to be just fine.

Compiled by the Innocuous-Sounding Institute for Common Sense Climate Solutions, with thanks to various funders that you really don't need to know about. ∎

Part B: The Solutions

Something's burning.

It could be kerosene, fueling a lantern that casts flickering shadows inside a fishing hut in Bangladesh; or gasoline in a small moped, kicking up dust as it takes a teacher to work in Tanzania. It's more likely to be diesel in a truck, delivering the latest load of special offers to a British supermarket; or natural gas, burning to heat a Canadian condo; or coal, turning turbines in a power station in the US, Italy or China.

Wherever it's happening, the result is the same: carbon, stored under the ground for millions of years in the form of fossil fuels, is combined with oxygen from the air to form carbon dioxide. One more tiny puff of this invisible, odorless gas drifts up from the hut or the schoolyard, the suburb or the industrial estate, to join the silent, gathering mass in the sky above.

Never in human history has there been such controversy, anger, protest, claim and counter-claim over a single chemical reaction. And with good reason: this chemical reaction – the burning of fossil carbon to produce carbon dioxide and heat – powers industrial society as we know it, and, as explained in Part A, could also change that society beyond all recognition.

Understanding climate change means understanding this connection. It's not enough to know about the science of greenhouse gases, or the greater risks of floods, droughts and storms in a warming world. We also need to understand the politics and the economics. We need to know why so little has been done to tackle the problem, even though we've been well aware of its magnitude for at least 30 years. We need to know what's driving this crisis, whose fault it is, and who's most likely to solve it. We need to know all about the different solutions on offer, but also who's proposing them, and why, and whether they really will tackle the root causes of the climate crisis.

Saying you know all about climate change because you understand the science is like declaring you're an expert on gardening because you've read a book on daffodils. The way that climate change interacts with people, with economies and cultures, with money and power – these are the vital things that we need to get our heads around if we're going to crack this problem.

The situation is serious. But there is still hope – so long as we act fast, and act well. Averting the worst of the climate crisis is about more than just cutting carbon – it's about envisioning new ways of living and working, of distributing power and of enjoying ourselves. It's about seeing beyond mainstream politics, and finding more powerful and exciting ways to make our voices heard. It's about seeking climate solutions based on the needs of people, rather than the desire for profit and growth. It's about bringing the people of the Majority World into the debate – the farmers, workers and indigenous communities who are at the frontline of the problem, and who may also hold many of the solutions.

These are challenging times – but they're exciting times too. We really do have the chance to change the world for the better. I hope the rest of this book will help you to figure out what part you'd like to play in making that change happen.

4 Where are all the emissions coming from?

The fuel with the most 'climate disaster points'... Coming clean in the global bathtub... The countries with the worst emissions... And those that should bear most responsibility.

Fossils and forests

The CO_2 from fossil fuel burning makes up the biggest part of our annual emissions and isn't too tricky to estimate. We have pretty good figures for the amount of oil, gas and coal that is mined and burned around the world, and we know how much CO_2 is produced from the burning of each of these fuels (see 'Fueling About', below). We can also estimate the amount of CO_2 emitted by cement manufacture, based on industry reports.

Another key source of emissions is 'land use change'. This innocent-sounding title covers a range of often quite destructive activities, the most important being the chopping down and burning of tropical rainforests. Measuring exactly how much CO_2 this releases is very difficult – it's not just the carbon dioxide released when the trees are burned; CO_2 is also released from the soil, and the removal of the forest affects the planet's ability to absorb carbon dioxide in the future. Forests are coming under new pressure from a growing demand for pastureland (thanks to rising global meat consumption) and the ongoing rush for biofuels (see Chapter 6).

Another important land use change is the loss of carbon from the soil via industrial agriculture practices such as chemical fertilization, plowing, and stubble burning;[1] peat-cutting and draining also leads to significant carbon emissions. The total amount lost in these ways is still not fully measured.[2]

Where are all the emissions coming from?

The Earth Policy Institute gives an estimated total for 'land use' CO_2 emissions in 2008 of 4.4 billion tonnes.[3]

Fertilizers, fields, floods, fridges and flights

The other main greenhouse gases that humans are responsible for – methane and nitrous oxide – come from such diffuse and hard-to-measure sources as landfill sites, fertilized fields, cattle herds, rice paddies, gas leaks and coal mines. Some estimates for these figures do exist, though, so I'm going to add them in.

We should also add in the available figures for the small-volume, high-impact 'fluoro' gases that we mentioned earlier. These have such enticing names as perfluoromethane, sulfur hexachloride and tetrafluoromethane. Some are produced for use as refrigerants or propellants, others are waste products from the manufacture of certain electrical products. They're only released in very small quantities, but have a lot of warming power per kilogram (you could remind yourself of this by looking back at the table on page 15).

One other thing to include is the extra warming power of aviation. In addition to our old friend CO_2, jet fuel burned at high altitude also produces a complex cocktail of other warming and cooling effects. The most recent research, looking at aviation emissions over the same 100-year period that is used to compare all other greenhouse gases, has rated airplane exhaust as 1.3 times more powerful than CO_2 alone.[4] Research into this is still ongoing, but this is the most recent figure so let's add it to our total.

The big dirty picture

The table on the facing page brings all of the above emissions sources together for comparison. It's pretty complex, so on page 74 are another table and pie chart that summarize the information.

Approximate breakdown of humanity's global greenhouse gas emissions for 2009

CO_2 Source	Billion tonnes of CO_2e/yr	%
Coal - Electricity and Heating	8.0	17.3%
Coal - Industrial Use	4.6	9.8%
Oil - Overland Transport	6.0	12.9%
Oil - Shipping	1.2	2.6%
Oil - Aviation	0.9	2.0%
Oil - Other	1.7	3.6%
Oil - Industrial Use	1.2	2.6%
Gas - Industrial Use	2.0	4.3%
Gas - Electricity Generation	1.8	3.8%
Gas - Heating	1.9	4.1%
Gas - Other	0.2	0.4%
Other Fuels	0.1	0.3%
Cement manufacture CO_2 (non-energy)	1.9	4.1%
Land use change CO_2	4.4	9.4%
N_2O from fertilizer use	2.4	5.2%
Methane and N_2O from livestock	2.5	5.4%
Methane from rice paddies	0.7	1.5%
Methane and N_2O from other agriculture	0.8	1.7%
Methane and N_2O from landfills and sewage	1.5	3.2%
Other greenhouse gases	0.5	1.0%
Methane and N_2O from fossil fuel use	2.3	5.0%
TOTAL	46.6	100.0%
Of which CO_2	35.7	77%
Of which other greenhouse gases	10.9	23%

Data compiled from International Energy Agency, BP Global Energy Review, Earth Policy Institute and Intertanko/IMO. Methane and nitrous oxide (N_2O) emissions are based on the last detailed breakdown that's publicly available (by the World Resources Institute in 2005), and assume that these emissions have risen at the same rate as CO_2 emissions from 2005-2009.

Where are all the emissions coming from?

A simplified breakdown of humanity's global greenhouse gas emissions for 2009

CO_2 Source	Billion tonnes of CO_2e/yr	%
CO_2 from Coal	12.6	27.1%
CO_2 from Oil	11.0	23.7%
Methane and N_2O from agriculture	6.4	13.7%
CO_2 from Natural Gas	5.8	12.6%
CO_2 from Land use change	4.4	9.4%
Methane and N_2O from fossil fuels	2.3	5.0%
CO_2 from Cement manufacture	1.9	4.1%
Methane and N_2O from waste	1.5	3.2%
Other greenhouse gases	0.5	1.0%
CO_2 from other fuels	0.1	0.3%
TOTAL	46.6	100.0%

A simplified breakdown of humanity's global greenhouse gas emissions for 2009

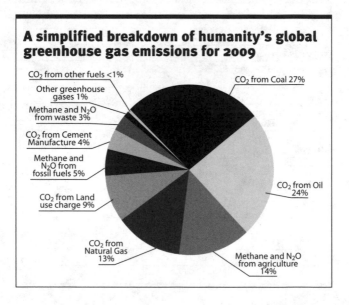

It's important to remember that these figures are all approximate. The CO_2 emissions from burning fossil fuels and making cement are pretty accurate (within 10 per cent, according to climate scientist James Hansen) but we've already noted that other greenhouse gases and land use change emissions are very hard to measure. Don't be surprised if you see slightly different versions of this breakdown (based on different measurement methods and dates) in other books, papers or websites.

There are two final sources of greenhouse gases that aren't included here:

• Major dams flood large areas of land, and the flooded vegetation then rots underwater, releasing methane. This means we can't really refer to big dams as 'low carbon' sources of electricity.[5] However, the precise amounts of methane released by dams have not yet been fully measured.

• Methane and carbon dioxide are being released from melting tundra and gradually heating soils as a result of the global warming we've caused so far. Only rough and partial estimates exist for these figures.

Here are some important points to take away from this whirlwind of numbers and charts:

• Fossil fuels are responsible for about 70 per cent of our annual greenhouse gas emissions. Any serious solution to climate change has got to involve a reduction in fossil fuel use.

• Emissions are widely spread across different global activities – they're linked to transport, homes, industry, agriculture, construction, waste, and pretty much everything that uses electricity or heating. There's not going to be a single magical solution to all this – we're going to need changes across the whole of society.

• Methane and nitrous oxide make up more than 20% of our current annual emissions.

This third point may be a bit misleading, because the standard way of comparing different greenhouse gases is based on their impact over 100 years. Methane has a high impact but doesn't stay in the atmosphere long, which means that it's rather more important in the short term than is revealed by the standard global warming calculations. It's like the difference between an hour on a sunbed and a minute with a blowtorch – they might deliver the same amount of heat in total, but the latter has a much more noticeable short-term impact. We'll need to bear this in mind when we're looking at emission reduction targets in the next chapter.

Something else about these numbers that slaps you straight in the face is the role of coal. It's the biggest part of our annual emissions. This isn't just because we burn so much of the stuff, but because some fossil fuels are inherently more polluting than others…

Fueling around

'*Shell is a carefully balanced blend of aromatics, naphthenes, and a small quantity of volatile paraffin hydrocarbons!*' – *1920s magazine advert for Shell Oil*[6]

Ah, fossil fuels – the commodity we love to hate. We know they're polluting, we know we should be shifting to alternatives, and yet they fill our lives. The goods and services we take for granted are coated in an oily rainbow sheen, covered with a dirty dusting of coal, and carry the distinctive whiff of natural gas. Usually we don't notice; perhaps when we're filling up our cars, switching on a gas hob, or shoveling coals onto a barbecue we might spare a thought for the fossilized carbon we're transforming into climate-wrecking greenhouse gases, but what about when we're sending an email, buying a pair of jeans or grabbing a dodgy late-night kebab? All of these activities will have been powered by fossil fuels somewhere along the line.

Fossil fuels play such a key role in the climate change story, it's worth looking at each of them in a bit more detail. One important attribute that we need to compare is: how much CO_2 do they produce?

The more carbon per kilogram each fuel contains, the more CO_2 will be produced when it is burned. To compare this fairly, we need to consider the amount of CO_2 that is emitted for every unit of energy we derive from burning that fuel. The energy unit we'll use is Kilowatt-hours (KWh) – a unit used on many home energy bills. This means that each type of fuel will end up with a score measured in grams of CO_2 equivalent per KWh (gCO_2e/KWh), which I also like to call Climate Disaster Points.* We're going to include the average emissions from extracting, transporting and burning each fuel, to try to arrive at the most complete picture that we can.[7]

This is a book about climate change. However, we noted in the introduction that we should also consider other global issues –social justice, human rights, war and conflict – at the same time. With this in mind, we couldn't possibly look at the carbon emissions from fossil fuels without considering some of their other impacts too. These are included in the profiles below (which are written from the perspective of the fossil fuels themselves).

Name: Coal

Climate disaster potential: Let's get one thing clear: I'm the major climate culprit here. I'm so heavily packed with carbon that every KWh of energy I provide also churns out 360 gCO_2e/KWh (including mining, transportation and methane leaks), which is far higher than oil or gas can manage. I don't only supply more than a quarter of humanity's current annual emissions – I'm also responsible for about a third of all the CO_2 ever

---

* Note that when these fuels are burned to make electricity, lots of energy is lost in the process and so the Climate Disaster Points per KWh of final energy becomes much higher. For example, electricity from coal is responsible for around 1,000 gCO_2e/KWh.

emitted by humankind, most of which is still in the air.

Main uses: These days, I'm all about electricity generation and industrial heat – for example, for making cement and steel. I currently provide about 41 per cent of the world's electricity. It's amazing how I've managed to hang on in there for so long, despite being the filthiest of all fossil fuels – I guess I'm just too cheap and abundant to leave behind.

Amount used in 2009: 13,000,000,000 tonnes, or 38 trillion KWh.

Electricity generation: 64%

Industrial use: 36%

CO_2 emissions in 2009: About 13 billion tonnes.

Local destruction: I have an extraordinary history of wilderness being ripped up for opencast mines and millions of people sent to their deaths down coal pits (explosions, cave-ins or lung diseases). Smog, air pollution, radioactive waste (yes, really), the destruction of entire landscapes through 'mountaintop removal' – I've got it all.

War and conflict: OK, I don't score so highly here – I'm just too abundant and well spread-out. China, the US, India, Australia, Indonesia, South Africa, Russia... they all have huge reserves. Of course, there are plenty of local struggles, with communities fighting local pollution and unwanted coal developments, but there just aren't enough guns involved to match oil's impressive conflict record.

Future prospects: Don't worry – there's more than enough of me out there to trigger runaway climate change all by myself. You'll get there sooner or later. Just keep building the coal plants and digging the mines.

Name: Natural Gas

Climate disaster potential: I am relatively light on carbon, and so produce only 220 gCO_2e/KWh (including drilling, transportation, and methane leaks).

This makes me the least polluting of all the fossil fuels per unit of energy – but don't be fooled. I've still got plenty of disaster potential.

Main uses: As well as being used in factories and offices, hundreds of millions of homes rely on piped-in gas for heating and cooking – something that can't be changed easily, so I'm well in there. Many countries have also turned to me as a key source of electricity generation over the last few decades – although I've not been able to overtake coal.

Amount used in 2009: 3,000,000,000,000 cubic meters, or 31 trillion KWh.

Industrial use: 34%

Electricity generation: 30%

Residential heating: 20%

Commercial heating: 13%

Other: 3%

CO_2 emissions in 2009: Around six billion tonnes.

Local destruction: I tend to be found alongside oil, and so share many of its impacts. Rather than spills, the major risk from me is that of explosions, like the one in Chongqing, China, in 2003 that killed 233 people, or the more recent ones in the US (Texas, Pennsylvania, and West Virginia) in 2010.

War and conflict: In both 2009 and 2010 Russia (one of the world's biggest gas producers) decided to cut off the gas supply to several eastern European countries due to ongoing political disputes.[8] When BP wanted to build a gas terminal in West Papua, they had to work in partnership with the Indonesian government, lending legitimacy to its illegal (and brutal) occupation of the country.[9]

Future prospects: Things are looking good for me, thanks to new techniques called 'horizontal drilling' and 'hydraulic fracturing' that allow energy companies to get me out of previously inaccessible shale rocks. The industry is estimating there's up to 100 years of extra gas out there.[10] Great news for me, and terrible news for the climate, which is set to be changed beyond recognition just by burning the fossil fuels that have already been found. Plus, gas shale extraction has such a high level of methane leaks that this 'new gas' may as bad for the climate as coal in the short term,[11] not to mention the reports of contaminated drinking water at my new hydraulic fracturing sites.[12] I told you I still had potential...

Name: Crude Oil
Climate disaster potential: Burning me produces around 325 gCO_2e/KWh, including drilling, transportation, and methane leaks. That's just for 'conventional' oil drilling; if you extract me from oil shale or tar sands you can count me at more than 400 gCO_2e/KWh because of all the extra energy required to extract and refine me. Which beats coal.
Main uses: I'm a liquid. This makes me incredibly useful in loads of situations, particularly, of course, for fueling vehicles. I'm also the main ingredient in most artificial fertilizers, plastics, and many other chemicals. I'm the fossil fuel that gets everywhere – imagine running the world's transport systems without gasoline and diesel, or industrial farms without fertilizers, or hospitals without plastics. Humanity's totally hooked on me – even if they do shift away from coal, I reckon I'll still be around to plunge them into climate chaos.
Amount burned in 2009: 4,000,000,000 tonnes, or 52 trillion KWh.
Transport: 61%
Non-energy use (plastics, fertilizers, chemicals): 16%
Industrial use: 10%
Other uses: 13%
CO_2 emissions in 2009: Eleven billion tonnes.
Local destruction: Around 84 million barrels of me are transported around the world every day, which means there are always going to be oil spills, devastating land and ocean whenever they occur. Then, of course,

you have all my refineries, gas flaring and other oil industry practices that routinely pollute the air, soils and waters in many countries. The 2010 spill in the Gulf of Mexico garnered headlines due to its size and the fact that the US was at risk – but far more oil has been poured into the soils and waters of countries like Nigeria and Ecuador over the years, without the rest of the world paying much attention.[13] Meanwhile, the extraction of oil from the Canadian tar sands is one of the world's most highly criticized projects, destroying vast swathes of ancient forest and poisoning the lands and waters of the indigenous people of the area.[14]

War and conflict: I'm an incredibly valuable resource concentrated in a relatively small number of locations. It's not very surprising that I've so often been a source of conflict – look at the recent histories of Iraq, Sudan, or Nigeria for some stark examples. It's not just the obvious wars, either – why do you think the West overthrew the popular Iranian government and reinstalled the brutal Shah in 1953? Why do you think certain oppressive regimes are spurned by wealthy governments, while others (such as Saudi Arabia) are welcomed into trade deals with open arms?

Future prospects: We've already found all my easy-to-reach reserves – which is why oil companies are turning to deep ocean drilling, newly revealed areas of the Arctic (thanks, global warming!) and 'unconventional' oils such as oil shale and tar sands to try to maintain their supplies. As Fatih Birol, Chief Economist of the International Energy Agency put it in November 2009, 'the era of cheap oil is over'.[15]

But – this is the good bit – we've barely scratched the surface of those unconventional oils! The tar sands extraction project in Canada is already the world's biggest industrial development – with toxic waste ponds so vast that they're visible from space – but it's the biggest oil reserve outside Saudi Arabia, so there's still plenty of me to be dug up there. Meanwhile, my oil company friends have found tar sands reserves in Congo-Brazzaville, Madagascar, Jordan... even if humanity manages to shift away from coal, there's more than enough of me out there to cook the planet, and I'm the one they're *really* addicted to. So don't listen to that has-been coal – I'm the main climate villain here.

Carbon world

The next question is: who's most to blame for all this fossil fuelery? The graph below shows CO_2 emissions since 1990, broken down between major groups of countries. It shows CO_2 only, and does not include land use change, so represents about 70 per cent of humanity's greenhouse gas emissions.

It seems that a seven-per-cent drop in CO_2 emissions from the industrialized nations – due to the financial crisis and relatively high oil prices – has been counterbalanced by an ongoing rise in CO_2 emissions from the 'developing' nations, particularly China (up

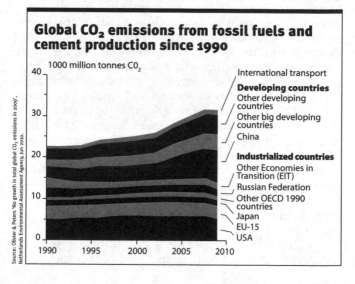

Global CO_2 emissions from fossil fuels and cement production since 1990

Source: Ollivier & Peters 'No growth in total global CO_2 emissions in 2009', Netherlands Environmental Assessment Agency, Jun 2010.

nine per cent) and India (up six per cent). This has caused global CO_2 emissions to flatten off in 2009.* Emissions from the rich world were on the rise again in 2010, so this looks like just a temporary blip, but the international recession may have bought us a little more time to deal with the climate crisis.

The graph also shows that, around 2007, China overtook the US to become the world's biggest emitter, while other large developing nations (Brazil, India, Mexico, Indonesia) have also greatly increased their global share this century. However, this doesn't give us the full picture, because all these countries are different sizes with different populations. China has just over four times as many inhabitants as the US; India has about ten times as many people as Japan.

Also, many of the emissions from Southern countries are associated with providing goods for the rich North. Factories in China, India, Taiwan and Indonesia are churning out plastic toys and electronic gadgets for Europe, Australia and North America. A recent study found that a third of China's emissions were associated with export goods. The UK's emissions may be around 35 per cent higher than reported once imported goods are taken into account.

It therefore isn't fair to directly compare the emissions from, say, the US and China. Imagine the US as one well-off household, while China is one wealthy family plus three poorer ones. The poor families spend a lot of their time driving around on errands for both the rich families, and racking up large home energy bills from doing their wealthy neighbors' laundry and cooking. It's not surprising that the four 'China' households in this scenario have a higher total carbon footprint than the single 'US' household. This illustrates how simply dividing up

---

* This is the first time annual global emissions have leveled off since 1992 (which was due to the collapse of the USSR and the closure of many European coal power stations in favor of gas-fired replacements).

emissions by geographic boundaries doesn't tell us the full story.

We can make things considerably fairer by working out each country's CO_2 emissions on a per capita basis. The table below and charts opposite show both the total and the per-person annual CO_2 emissions of the 20 highest-emitting countries. These 20 countries are responsible for about 80 per cent of global carbon dioxide emissions from cement and fossil fuels.

Total and per-person annual CO_2 emissions (from fossil fuels and cement production) of the 20 highest-emitting countries in 2009.

	Annual Emissions (Millions of tonnes of CO_2)	Percentage of global CO_2 emissions	Tonnes of CO_2 per person in 2009
China	8,060	26.0%	6.1
USA	5,310	17.1%	17.2
India	1,670	5.4%	1.4
Russia	1,570	5.1%	11.2
Japan	1,180	3.8%	9.2
Germany	770	2.5%	9.3
Iran	570	1.8%	7.7
South Korea	560	1.8%	11.5
Canada	540	1.7%	16.3
UK	490	1.6%	8.1
Mexico	470	1.5%	4.2
Indonesia	440	1.4%	1.9
Italy	410	1.3%	7.0
Australia	400	1.3%	18.8
Brazil	380	1.2%	1.9
South Africa	380	1.2%	8.0
Saudi Arabia	370	1.2%	13.6
France	370	1.2%	6.0
Spain	310	1.0%	7.1
Ukraine	310	1.0%	8.0
TOTAL	24,560	79.2%	n/a

Annual CO$_2$ emissions from fossil fuels and cement production of the 20 highest-emitting countries in 2009.

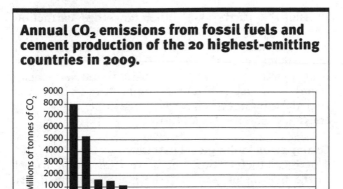

CO$_2$ emissions per capita, from fossil fuels and cement production, in the 20 highest-emitting countries in 2009.

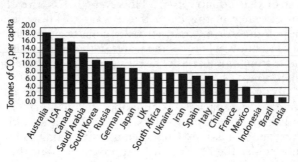

The bottom chart tells a rather different story from the previous two. Every person in the US is responsible, on average, for almost three times as much CO$_2$ as a person in China. Canada has very high per capita

Where are all the emissions coming from?

emissions – partly due to its tar sands extraction projects, which burn huge amounts of natural gas per year. Australia doesn't have a huge population but burns a lot of coal. The oil-rich nations like Saudi Arabia tend to have small populations and very cheap energy – in fact, Qatar, Bahrain, Kuwait and the United Arab Emirates have even higher per capita CO_2 emissions than the US and Australia.*

Coming clean in the global bathtub

This is all useful stuff, but if we *really* want to get to the bottom of who has the biggest responsibility for climate change we need to go back a bit further in time. What about all the CO_2 that's already been released? Carbon dioxide can stay in the atmosphere for up to 200 years, and it's the total amount that's up there that we need to be worried about.

Think of the atmosphere as being like a giant bathtub that's filling up with water (that's the CO_2), from a number of different taps/faucets (those are the various countries). The annual CO_2 emissions of each country tell us how fast the water is currently coming out of that country's tap. However, there's already a lot of water in the bath. If we want to know who's responsible for our current risky climate situation, we need to know where all the water in the bath originally came from, not just how fast it's coming out of the taps right now.

To figure out who's put the most water in the bath, we can trace back CO_2 emissions to the start of the Industrial Revolution – let's say 1850 – and add up everyone's emissions from then up until today. A lot of those older figures are fairly approximate, but they're good enough to give us a rough picture – with some very interesting results. The chart opposite shows the

---

* However, the *total* annual emissions from these oil-producing nations aren't high enough to put them in the top 20 global emitters, so they don't appear in the chart at the bottom of page 85.

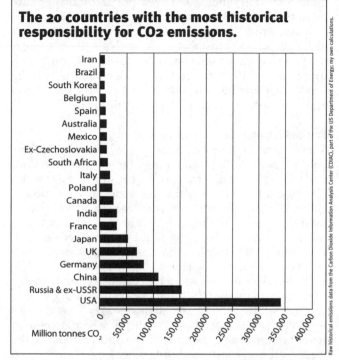

The 20 countries with the most historical responsibility for CO2 emissions.

Million tonnes CO_2

Raw historical emissions data from the Carbon Dioxide Information Analysis Center (CDIAC), part of the US Department of Energy; my own calculations.

top 20 countries with the most historical responsibility for CO_2 emissions (between 1850 and 2007).

This chart shows us quite a different picture. The US has put over three times as much CO_2 into the air as China. The UK is responsible for 1 in every 20 (fossil and cement) CO_2 molecules that have gone into the atmosphere. France has put more fossil carbon into the air than the whole of India.

This is because the big Southern nations have arrived rather late to the fossil fuel party. To go back to the bathtub analogy, China and India have only started really turning their bathtaps on in the last 30 years. The European nations already had their carbon taps

Where are all the emissions coming from?

on in Victorian times, and have been gradually making them flow faster ever since.

Russia and the Baltic states tend not to come out well from these kinds of historic analyses. Even though their emissions peaked in the late 1980s, and then dropped for 10 years following the collapse of the USSR, they still burned a lot of coal in their peak years and much of the resulting CO_2 is still in the air.

You're history

We're going to have one final play with these numbers before we start talking about how far we need to reduce our CO_2. There's one more thing I want to show you.

Those of us in wealthy countries owe a decent chunk of our current comfort to the fossil fuel energy used in the past. Without all that historic gas, coal and oil, we would not have the homes, transport systems, food supply, schools, hospitals, universities, emergency services and everything else we benefit from today. The early use of fossil fuels also gave certain nations a major advantage in the last 200 years of wars, conquests and empire-building that have shaped today's world. It's no coincidence that those countries which exploited fossil fuels first and fastest tend to be those that are the wealthiest and most powerful today.*

I therefore think it is fair to divide up the historical emissions of each nation by the number of current residents, to get a 'historical per capita responsibility' total. This is the amount of CO_2 that has been emitted since 1850 to provide each citizen of that nation with the lifestyle they currently enjoy.** This produces some very telling results, as can be seen in

---

* Although this is generally true, there are exceptions – for example, Eastern European and ex-Soviet states have burned a lot of fossil fuel in the past, but historical events have not left them with the same standard of living as Western Europe or the US today.

** This exercise only really works for countries that have made a significant contribution to global CO_2 emissions. I've therefore only included countries that have emitted at least 1% of all that historical CO_2.

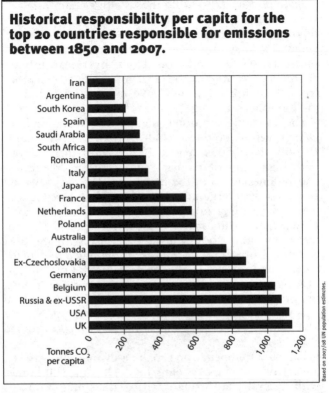

Historical responsibility per capita for the top 20 countries responsible for emissions between 1850 and 2007.

Iran
Argentina
South Korea
Spain
Saudi Arabia
South Africa
Romania
Italy
Japan
France
Netherlands
Poland
Australia
Canada
Ex-Czechoslovakia
Germany
Belgium
Russia & ex-USSR
USA
UK

0 200 400 600 800 1,000 1,200

Tonnes CO_2 per capita

Based on 2007/08 UN population estimates.

the chart above.

This tells a whole new story. Per resident, the UK, the US, Russia, Belgium and Germany have the largest historical responsibility for CO_2 emissions, at around 1,000 tonnes per person living today. Europe and North America dominate the chart, and India and China don't even make it into the top 20.

More emission omissions
There are a few more details we need to complete the global emissions picture, many of which pile yet more

climate responsibility onto the wealthier nations:

• These figures only include the emissions officially reported by governments. There are some 'dark' emissions that don't make it onto the books – particularly the pollution from overseas military action. For example, the unreported CO_2 emissions from US military fuel use have been estimated at 73 million tonnes a year.[16]

• The above figures don't include CO_2 from land use change; neither do they include methane, nitrous oxide, or other greenhouse gases. If they did, then nations with high rates of tropical deforestation, like Brazil and Indonesia, would be higher up the list, as would countries with large amounts of fossil fuel extraction, livestock farming and/or heavily fertilized cropland. However, as with fossil CO_2, the responsibility for these emissions doesn't always lie just with the host country. Tropical rainforests are being felled to provide pastureland for exportable beef. Industrial plantations in Africa, coated with nitrous-oxide-spewing fertilizers, provide crops for Northern markets. Gas leaks and flares are added to the national emissions of Nigeria and Kuwait, even though most of their oil is consumed elsewhere.

• The bald numbers don't make a distinction between 'necessity' and 'luxury' emissions. Heating 100 homes with natural gas through a Russian winter produces the same amount of greenhouse gas emissions as 15 return flights between London and New York. They each cause the same amount of global warming, but contribute different amounts of benefit to society.

• All the tables and graphs in this chapter also hide the fact that it's the better-off people in every country who are burning the most fossil fuel. This relatively wealthy minority are also consuming the largest amount of material goods, and so contributing the most to deforestation, agriculture emissions, and the CO_2 from those Chinese and Indian factories.[17]

You've probably got the point by now. Current greenhouse gas emission rates are useful for figuring out what solutions are needed where – but they don't tell us the full story. The wealthy Northern nations bear the biggest historical responsibility for bringing the planet to the brink of climate crisis, on a number of different levels. Newly fledged high-emitters like China, India and Brazil urgently need to shift onto a cleaner development path. However, it's only fair that the Northern countries – who have reaped the greatest benefits from the last 200 years of emissions – should take the lead, and provide serious financial and technical support to help the rest of the world make the shift to a low-emission society.

1 See, for example, an interview with a soil scientist at http://nin.tl/dBIMrl 2 Natural England (2010) http://nin.tl/bCKe2w 3 Earth Policy Institute (2010), http://nin.tl/bEcDkb 4 Earlier figures were 2.7 times (from the IPCC in 1997) and 1.9 times (from a European study called 'TRADE-OFF' in 2000). These figures are still commonly used by campaigners but are now rather out of date. The latest figures are from Forster, Shine and Stuber, *Atmospheric Environment*, 40, 6, 2006, plus an update (called a 'Corrigendum') by the same authors in *Atmospheric Environment* 41, 18. 5 DelSontro et al, 'Extreme Methane Emissions from a Swiss Hydropower Reservoir: Contribution from Bubbling Sediments', *Environ Sci Technol*, 44 (7), 2010. See also International Rivers (2007), www.internationalrivers.org/node/1398 6 As seen on www.advertisingarchives.co.uk 7 The CO_2e figures in this chapter are calculated using the following sources: D Weisser, *Energy* 32, 9, Sep 2007, www.iaea.or.at ; B Sovacool, *Energy Policy* 36, 2008, available at www.nirs.org ; DOE/NETL, 'Consideration of Crude Oil Source in Evaluating Transportation Fuel GHG Emissions', 2009, www.netl.doe.gov ; R Howarth, 'Preliminary Assessment of the Greenhouse Gas Emissions from Natural Gas obtained by Hydraulic Fracturing', Cornell University, 2010, www.eeb.cornell.edu 8 Associated Press, 'Pleas but no progress in European gas crisis', 14 Jan 2009. 9 See, for example, http://dte.gn.apc.org/73tan.htm and www.free-westpapua.org 10 Nolan Hart (2010), http://nin.tl/csqyoN 11 R Howarth, (2010) 'Preliminary Assessment of the Greenhouse Gas Emissions from Natural Gas obtained by Hydraulic Fracturing', Cornell University, 2010. Available at www.eeb.cornell.edu 12 'Dash for Gas Raises Environmental Worries', *New York Times*, 12 Jul 2010. 13 'Gulf oil disaster not unique to BP and will happen again', *The Ecologist*, 9 Sep 2010. 14 Indigenous Environmental Network, www.ienearth.org/tarsands.html 15 BBC news, 10 Nov 2009, http://nin.tl/bTbjS1 16 B Sanders, *The Green Zone*, AK Press, 2009. 17 See, for example, 'Most polluting postcodes in Britain identified in heart of middle England', *Daily Telegraph*, 19 Jul 2009.

5 How much do we need to cut?

Why setting targets for cutting emissions will not do the trick... The minimum 'red line'... Leaving fossil fuels in the ground... Why solutions must be just – or they will bust.

SO WHERE ARE we trying to get to? What is a 'safe' level of global CO_2 emissions?

You may remember from Chapter 1 that the amount of CO_2 in the air currently stands at around 388 parts per million (ppm). This is the key number to focus on – in our bathtub analogy, this ppm figure represents the amount of water in the tub. If you add in the other greenhouse gases, the total is more like 435 ppm CO_2e – but we're just going to home in on that 388 ppm of CO_2 for the moment.

When governments forged an international climate agreement at Kyoto in 1997 (see Chapter 6), an upper CO_2 limit of 450 ppm was generally agreed upon as a useful target. Stabilizing CO_2 at this level would, it was believed, cause a temperature rise of 2°C, which was considered at the time to be manageable. However, more recent evidence – particularly relating to the risk of feedback loops, melting ice sheets and runaway warming – suggests that 450 ppm is in fact a dangerous level of atmospheric CO_2 and 2°C an extremely risky level of warming.

A more useful target, according to a growing body of scientists, including leading NASA scientist James Hansen and IPCC Chair Rajendra Pachauri, would be 350 ppm of CO_2. If the other greenhouse gases remain stable, 350 ppm of CO_2 is the level at which the planet will stop warming up.[1] The sooner we get atmospheric CO_2 down to this level, the more of those dangerous future risks we can prevent.

This means that, at 388 ppm, we're starting from too high a level of carbon dioxide in the air, and need to reduce it. Our carbon bathtub is already overflowing, and starting to wreck our bathroom. We need to get the water level down before it floods the whole house and the floors start collapsing. Unfortunately, the plughole is very small (the water going out of the plughole is the CO_2 that's being gradually sucked out of the atmosphere by oceans and plants). So long as we are putting water into the bath faster than it can drain out, that ppm figure will keep on rising.

Can we cut it?

Much of the public debate around targets has therefore focused on reducing our rate of emissions – the speed at which water comes out of the taps or faucets. The original Kyoto agreement aimed to hold CO_2 at 450 ppm by achieving a 60-per-cent reduction in global emissions by 2050, and individual countries were given national targets that were supposed to fit in with this goal. You've probably heard campaigners calling for more stringent global or national cuts since then, as the science has become more serious. There are advantages to this kind of approach: looking at percentage cuts gives us some idea of the kinds of changes we need to make in our society, and forces us to ask important questions. How could we heat our homes, travel around and grow our food while producing 60 per cent, or 80 per cent, or 90 per cent less CO_2?

Unfortunately, measuring the annual emissions of different countries, industries and communities is a highly complex business. A huge number of different, overlapping activities contribute to the annual emission levels of any nation. This means that it's relatively easy for politicians to fudge the numbers, pick different starting dates, argue endlessly about what should and should not be included, and blame all sorts of external factors whenever emissions rise. It also allows for the fiddling of the figures through dubious carbon offsetting and trading schemes (see Chapter 9). Setting long-term targets – up to 40 years away – also makes it easier for politicians and other decision-makers to put off taking serious action.

Meanwhile, more and more scientists are saying that emissions reduction targets aren't even that helpful from a scientific point of view. We should be talking about the total amount of greenhouse gas in the atmosphere, not just the annual speed at which we're emitting it. For example, if we let our emissions rise steadily until 2045, then had five years of massive cuts, we might just end up with (for example) an 80-per-cent emissions reduction

---

What about population?

Some people argue that population reduction is important for fighting climate change. It's true that the more people there are on the planet, the fewer resources there are to go around. However, birth rates in most Northern nations are low; most population growth is occurring in poorer countries. The current per capita consumption rate in these countries is very small – for example, the average Canadian uses the same amount of energy per year as 20 Tanzanians. The wealthiest 20 per cent of the world's people use over 70 per cent of the energy. With regard to climate change it is far more urgent to reduce consumption levels in the North than birth rates in the South.

High birth rates are strongly associated with poverty, hunger and a lack of access to healthcare. They are also connected to a lack of women's rights and restricted access to health information and con- traception. If we want the world's population to stabilize sooner rather than later, we need to support people around the world – especially women – in claiming more rights, greater dignity and full control over their lives. Unfortunately, many of the people who focus on population and climate change seem to prefer talking about top-down population quotas, migration controls and other restrictive and unjust measures. Population is often used as a distraction tactic or a form of denial, to avoid having to confront the far more urgent issue of cutting Northern overconsumption, and can have racist overtones ('If only *they* weren't having so many babies it would all be OK').

Population is not a special issue that we need to confront with any separate tactics. It's something that will solve itself if we can tackle poverty and hunger, and restore people's basic human rights around the world. Which – fair enough – isn't that simple, but many of the measures that will help us to tackle climate change (see Chapter 6) will also move us towards these goals. ∎

---

by 2050; however, if we think back to the bathtub we can see that this wouldn't be a helpful strategy. Running the water on a high setting for more than 30 years and then just turning it down sharply at the end, will still leave us with an overflowing bathtub. To have a decent chance at limiting the damage to our bathroom, we need to start turning the taps/faucets down now, not in 30 years' time.

On target

Once carbon dioxide is in the air, it stays there for a long time – potentially for hundreds of years. We also don't know how much more CO_2 plants and oceans can safely absorb – all that extra carbon dioxide is already making the seas more acidic, with knock-on effects for marine life. Even if the CO_2-sucking powers of nature remain intact, there's a limit to how much more CO_2 we can put into the atmosphere before that ppm figure gets so high that we've no chance of bringing it back down again in time to avoid runaway change.

This fact – that there's a limit to how much CO_2 we can put out there – means that we can calculate roughly how much more coal, oil and gas it will take to push us past the point of no return, or, to put it another way, how much of it we need to leave in the ground to avoid runaway warming.

James Hansen has helpfully calculated that if we phased out coal burning in the next 20 years or so, and only burned *existing* reserves of oil and gas (in other words, we stopped looking for new oil and gas fields through things like deep sea drilling and Arctic exploration), we could get atmospheric CO_2 down to 350 ppm in time to prevent runaway climate change. This comes with a few other caveats: we'd also need to stop burning 'unconventional' sources of oil and gas, like tar sands and shale, because these are highly polluting and would easily tip us over the edge. There'd

need to be separate action on 'land use' emissions –
we'd need to halt mass deforestation and to stop the
loss of carbon from the world's soils. Finally, we'd
need our emissions of other greenhouse gases to the
atmosphere (methane, nitrous oxide, 'fluoro' gases and
cement emissions) to start falling year-on-year.

This set of actions is, then, a kind of climate 'red
line' – the absolute minimum greenhouse gas reduction
that we need to make. If we don't achieve the things
in the paragraph above, then the state of the climate
will probably be out of our hands – we'll have a high
chance of being locked into feedback loops that will
heat the climate faster and faster, until we're left with
an unrecognizable planet.

Nothing to lose

This is obviously useful to know. But it's still not
enough. Because the climate crisis is already under
way, every single kilogram of greenhouse gas that
goes into the air is making things worse. If we accept
an arbitrary 'target' of a certain cut by a certain date,
we are effectively saying that a certain amount of
emissions is 'OK'. But at this point, *no* greenhouse
gas emissions are OK – they are leading directly to
death, destruction, and species loss around the world.
We need to get atmospheric CO_2 down as low as we
can, as soon as we can, in order to save as many lives,
livelihoods, homes and species as possible.* If we aim
for anything else we're essentially saying that a certain
level of death and destruction is acceptable. Even
though this position presents a significant political
challenge, it's the only moral option, and it also means
we've still got a chance of avoiding runaway change if
things turn out to be worse than expected.

So after all of that number-juggling, we're left with

---

* Grassroots activists at the Cochabamba alternative climate summit in 2010
called for a target of 300 ppm for this very reason.

one, simple, clear target: zero emissions, as fast as we can.

This means leaving fossil fuels in the ground – as much of them as possible. The sooner we can stop using them, the better. We've already seen how destructive they are, even without their effect on the climate. Every tonne of coal, every barrel of oil, every cubic meter of gas that stays underground saves lives.* It's as simple as that. It means halting deforestation and reducing methane, nitrous oxide and other greenhouse gas emissions too – but fossil fuels form the biggest part of the problem, and so we need to transition to a society that doesn't rely on these filthy and dangerous energy sources. The alternatives do exist – not just different technologies, but different and better ways of living, working, and organizing ourselves. We need to get on with putting them into place.

Hansen's targets give us a good baseline, something to keep in mind – if we cross this line, if we're still burning coal in 2030, if we keep looking for new oil, or if we keep digging up the tar sands, then we're in *really* big trouble. But we should aim for the biggest, fastest change possible.

We'll come back to how we reach these targets in Part C of the book. In the meantime, there's one other important goal we need to add to our list.

Just or bust

In Chapter 4, we saw that it is the wealthy industrialized nations who are most responsible for bringing the world to the brink of climate crisis. In fact, the whole process is doubly unfair, because it's overwhelmingly people in the poorer nations (the Majority World or Global South) who are suffering the worst effects of

---

* It should go without saying that we need to make these emission reductions in a way that doesn't endanger people – we can't just turn off everyone's heating in cold countries or stop manufacturing medical supplies. We need a rapid transition to zero-carbon alternatives, not a mass switch-off!

climate change. 95 per cent of the victims of natural disasters live in poorer nations.[2]

The Northern countries also bear a share of blame for increasing the vulnerability of many poor nations to climatic disasters, thanks to past colonial practices. Countries that were once extractive colonies are more likely to have high levels of inequality, restricted land rights, large rural populations and limited press freedom, which are all vital factors in increasing the number of deaths and lost homes when storms, floods and droughts strike their shores.[3] Think, for example, of recent weather-related disasters in Burma, Niger or Pakistan.

The wealthy nations, then, owe a serious debt to the Global South – our consumption of 'cheap' fossil energy has had a very high price for much of the rest of the world. Any solutions we put in place should include a strong element of reparation, and of justice.

This isn't just a matter of principle – any solutions which do not take into account the needs of the Majority World will be fought against bitterly by the people of these nations. This would be a disaster for us all. We too often think of climate solutions as something we in rich countries need to come up with, and then 'give' to the rest of the world. This just isn't true – as we'll see in Chapter 6, some of the most important solutions, particularly with regard to sustainable land use, are coming from the people of the South. While we in the North have a responsibility to curb our fossil fuel use, and to provide funds and appropriate technology to other nations to help repay our climate debt, we cannot and should not force a set of top-down 'solutions' from wealthy nations onto the rest of the world (however well-meaning they might be). It's not only morally wrong – it simply won't work.

If we want countries like China, India and Brazil to shift away from coal power and halt deforestation, then the real pressure will have to come from the people who live in those countries. If we want them to succeed, we'll need to work together with them and build a truly global movement for change. Grassroots social movements in the Global South, and 'frontline communities' being affected by climate change all around the world, are an incredibly important part of this picture. It won't be possible to stop global climate change without their efforts – and they will not want to work together with anyone who isn't taking the needs of their communities into account. A crucial issue for people in many Southern countries is the preservation of natural resources for the common good, and the protection of local people's sovereignty over land, energy, forests and water. Many of the 'solutions' being proposed by wealthy Northern governments – such as biofuels, carbon trading, large-scale technofixes or forest privatization (see Chapters 6 and 9) – directly conflict with these primary concerns.

We also need to remember the needs of workers. We're talking about a very rapid transition from a global economy based on fossil fuels to a safer, cleaner alternative. But millions of people all over the world work in the fossil-fuel industries. Many have little choice in the matter, because energy corporations have taken over their lands, communities and local economies and left them with few alternative methods of survival than to work extracting the fuels, or providing services to those who do. For example, the sudden rush for tar sands in Alberta, Canada, has led to huge increases in house prices and the cost of goods and services, forcing many residents to seek jobs in the industry just to make ends meet. Often the people suffering the most negative health impacts from fossil fuels are those who also work on their extraction, refinement and transport, such as the five million people who work in China's coal industry. Any effective global movement for change needs to include, not exclude, these workers. This means we want solutions that give workers in fossil-fuel

How much do we need to cut?

industries the power and opportunity to shift to less polluting livelihoods, in a way that works for them. This is often called a 'just transition' away from fossil fuels, and it's true for other industries too (aviation, intensive agriculture, manufacturers of cheap plastic goods, and so on).

Both targets and principles

In summary, here is our 'red line' minimum scenario. This is what we need to do to have a decent chance of stopping runaway global warming:

- Phase out coal burning by 2030
- Stop burning tar sands and oil/gas shale
- Stop exploring for new oil and gas, and burn only the oil and gas that we've already discovered (at the very most)
- Halt deforestation and the loss of carbon from soils
- Make sure our emissions of methane, nitrous oxide, CO_2 from cement production and other greenhouse gases are falling year-on-year

This is what we need to keep in the back of our minds as the absolute bare minimum, don't-even-go-there target. But day-to-day, we can sum up our emissions goals even more simply:

Keep the fossil fuels in the ground, halt deforestation and the loss of soil carbon and shift to a zero-emissions society as fast as we can.

Finally, we need to make sure our solutions are fair as well as effective. A term that's frequently used to describe this is 'climate justice'. Here is some text based on various declarations from Southern activists that should help to guide us in the right direction:

Ensure that our emissions targets are met in a fair way that recognizes the historical responsibility of industrialized Northern nations for the climate crisis. The solutions to the climate crisis should not have negative knock-on effects on the world's poor or the

fair.[1] Some of their zero-carbon plans rely on uncertain and dangerous technologies such as carbon capture, biofuels or nuclear power (see the boxes below, which explain why none of these things are likely to help us). Some are based on unequal access to energy amongst the world's people. And some put faith in massive transnational projects like solar mega-farms in the world's deserts linked to huge regional electricity networks. Negotiating, financing, and building these kinds of cross-border megaprojects without placing huge power in the hands of a small number of nations, or steamrollering over the needs of local people, would be an even greater political challenge than the already-difficult task of getting governments to build renewable generation within their own borders.

A more general problem with energy plans on this scale is that they risk coming up against serious limits: limited funds, limited energy, limited mineral resources, political barriers or – most important of all – limited time. We need to have cut out most of our fossil-fuel use by 2030, and the less new infrastructure we have to cram into that time, the better our chances will be. Finally, like it or not, we'll need to use fossil fuels to build at least some of our new zero carbon world, and that means one final burst of carbon emissions, which we must keep as small as possible – it would be a terrible irony if this final surge tipped us over the edge into runaway climate change.

Lightening the load

Luckily, there is another way. Researchers at the Centre for Alternative Technology in Wales have shown that it's perfectly possible to live a comfortable Northern lifestyle on just 15,000 KWh of energy per person per year.[2] That's a significant reduction for the industrialized nations, but a jump upwards for most of the world.

The chart opposite shows how this kind of shift

planet's ecosystems. The transition away from fossil fuels should be a just one that respects the rights of workers in relevant industries. Natural resources should be seen as existing for the common good, and local people's sovereignty over land, energy, forests and water should be protected.

Now it's time to test the various climate solutions on offer against these targets, and see how they stack up.

1 Hansen et al, *Open Atmos Sci J*, vol 2, 2008; Rockström et al, *Nature* 461, 2009.
2 J Timmons Roberts and Bradley C Parks, *A Climate of Injustice*, MIT Press, 2007. **3** Ibid.

6 What are the solutions?

How to live comfortably on a fair share of carbon... Transport choices... Why biofuels, carbon capture, nuclear power and technofixes won't help... The renewable energy sources of the future... A better world is possible.

WE NOW HAVE an idea of the scale of change needed to avoid runaway warming. It's pretty big. This raises the important question: are we screwed? We can't just switch off all the power stations, cars and central heating in the North, and we can't expect the majority of the world to remain in poverty. Is it physically possible for everyone in the world to enjoy the lifestyles that so many in the North now take for granted – safe and comfortable homes, labor-saving appliances, easy travel, healthcare, education, plentiful food – while giving up the fossil fuels?

You'll be pleased to hear that yes, it certainly is possible. But it won't be easy. A fair and effective route to a post-fossil future will require all of the following:

- The industrialized world must reduce its per-capita energy use whilst maintaining a good standard of living
- The rest of the world must be given the same access to energy, and the poorest lifted out of poverty, using zero-carbon methods, policies and technologies
- We need to roll out sustainable technology on a massive scale, to replace fossil fuel infrastructure across the globe.

Let's look at each of these points in turn.

A rising energy tide

Most of the world's energy is consumed in a relatively few countries – the US alone is responsible for 27 per cent, the EU for 22 per cent, and Russia, Japan, Canada and Australia together make up another 19 per cent. That means 68 per cent of global energy is consumed by just 16 per cent of the population. China uses slightly more energy than the US in total, but much less per person. The table below shows some approximate per-capita energy use figures for different countries:

Approximate energy use per person in selected countries and regions	
Country/Region	Energy Use Per Capita (KWh in 2008)
Canada	96,000
USA	89,000
Australia	75,000
EU	48,000
China	19,000
Average for Global South	5,500
Tanzania	4,000
Nepal	3,500

Source: International Energy Agency (IEA), 2008

Some 85 per cent of all this energy use currently comes from fossil fuels.

Not everyone in the world aspires to a Northern lifestyle; nonetheless, a fair solution to climate change should offer everyone on the planet a similar amount of energy and resources, even though many millions of people would probably choose to use less than their share. As we noted in Chapter 5, keeping justice at the heart of our climate solutions will be vital for building the powerful global movement we need.

So let's imagine that we're going to bring everyone in the world up to an EU level of energy use. Could we manage that on renewables alone? Even if the wealthy world holds its energy use steady, this will still be an incredibly difficult task. A number of energy analysts have tried to tackle this challenge and now have come up with a solution that's both reliable

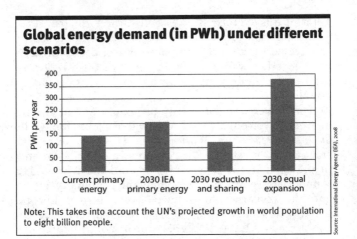

Global energy demand (in PWh) under different scenarios

PWh per year

400
350
300
250
200
150
100
50
0

Current primary energy | 2030 IEA primary energy | 2030 reduction and sharing | 2030 equal expansion

Note: This takes into account the UN's projected growth in world population to eight billion people.

Source: International Energy Agency (IEA), 2008

in energy use would compare to today's energy use, and to the International Energy Agency's business-as-usual prediction; they believe that global primary energy use will rise from 145 to 201 PWh of primary energy (and 132 PWh of final energy, after electricity generation losses) by 2030.* Most of this energy would still come from fossil fuels and biofuels, and we'd be pretty much locked into the worst-case scary climate disaster scenarios. The final bar in the chart shows the amount of energy we'd need if the rich nations held their energy use steady while the rest of the world increased their per capita energy use up to European levels.

Don't get too hung up on the precise numbers here – they're all approximate, and I'll be the first to admit that reaching a totally zero-carbon and fairly shared energy scenario by 2030 is a teensy bit ambitious. The

---

* One PWh = one trillion KWh. Primary global energy consumption is the total amount of energy consumed by humanity each year. Final global energy consumption is the amount of energy that we actually use in our lives. The reason it's lower is because a large chunk is lost every year in generating electricity from fossil fuels, vanishing as heat up the power station chimneys.

What are the solutions?

point of this exercise is to establish three things:

1) It is possible – in theory – for everyone in the world to have access to a low-impact version of a Northern lifestyle if they want one, while simultaneously *reducing* total global energy use, despite an increasing global population.

2) This lower level of energy use could potentially be met entirely from existing renewable technologies. In *Sustainable Energy Without the Hot Air* respected energy expert David Mackay made some estimates of the capacity of renewables around the world.[3] He found that (with a huge renewables building effort and the construction of large solar farms) Europe could produce over 30,000 KWh/person/year. The US could reach 22,500 KWh/person on wind, hydro and geothermal alone; adding in solar would boost this even further. The global average works out at 10,000 KWh/person for non-solar technologies. Solar water heating, solar photovoltaics (PV) and concentrating solar power could then provide far more than the remaining 5,000 KWh per person.[4]

3) Providing fair access to energy without reducing Northern consumption would require *three times* as much energy, an impossible task with existing technologies.

For a decent chance at a safer future we therefore need a Northern energy reduction and a fairer sharing of global resources. This runs counter to the way our global economy currently works – we'll look at how to deal with this in later chapters.

We'll need to rebalance our energy use *within* countries too. Many nations will need to reduce energy use in some areas while letting it increase – in a low-carbon way – in others. This will be particularly important in countries like India and China, where a sizeable minority of their populations live relatively energy-intensive lives, but the majority still live at the low end of the energy-use scale.

Now we'll look at some of the practical solutions we could use to create this fairly shared, lower-energy, fossil-free scenario. Some of these solutions aim to reduce energy demand, making a high-quality life possible on just 15,000 KWh/year. Some offer alternative energy sources to replace fossil fuels, while others aim to cut emissions from agriculture and deforestation. These solutions will require a combination of sustainable technology and gradual public behavior change to achieve. Remember, this isn't meant to be a definitive blueprint – we'll probably come up with more and better solutions to add to this list along the way. This is just a demonstration that a low-carbon, and ultimately zero-carbon world, is already both possible and desirable.

Getting around

Car culture will need to be challenged. Mass car ownership in Northern countries is taken for granted, but is in fact a bizarre (and expensive, and stressful) extravagance. The vast majority of car journeys are repetitive commuter trips which could easily be replaced by cycling, walking or public transport. Bicycles are amazing devices, and our streets need a serious redesign to make room for them. If only we weren't so hung up on each having a car of our own, then a car-sharing system could easily meet most people's needs – we'd grab a car when we needed it for day trips, holidays and moving stuff around, and get about in more sensible ways the rest of the time.

Public transport needs proper public investment in order to be comfortable, reliable, well-planned and affordable. We need to phase in electric buses and trains, at the same time as we switch our electricity generation to renewables. A decent public transport system has enormous social benefits well beyond the climate issue. Less than 10 per cent of the global population own a car. Globally, 1.2 million people

What are the solutions?

were killed in traffic accidents in 2002, with as many as 50 million injured;[5] on current trends, the World Health Organization expects road accidents to kill more people than AIDS by 2020.

Freight can be reduced by cutting down on unnecessary consumption and goods transport (see below). We can also shift it onto electrified rail and waterways.

High-speed flying sadly requires so much energy that we're going to need to do a lot less of it in our zero-carbon future. The good news is that this may be easier than we think – most of the flying, even within Northern nations, is done by a wealthy minority of people. A person traveling by air from a UK airport has an average annual income of over $75,000. Half of the population of the UK do not fly in any given year.[6]

For many overland journeys, high-speed trains are already faster than flying once check-in time is included. For longer journeys, Northern populations will need to get used to traveling more slowly, by land or by sea – but this can lead to higher-quality holidays, and can help us to develop a better understanding of the world around us. It could also become easier to do in a less frantic economy, with less material consumption, less work and more leisure time.

Electric cars could either help or hinder us. Filling the world with electric cars powered by coal-fired electricity would actually increase emissions. However, if we succeed in significantly cutting car use in Northern nations and building sustainable transport systems in the South then we should have enough renewable electricity to power all the cars we need.

Hydrogen isn't a fuel, because it needs to be generated using electricity. It's a less efficient way of storing energy than charging up a battery, but it could be useful for powering long distance overseas transport, in cases when a battery wouldn't last long enough. Unfortunately, it wouldn't work as a fuel for high-speed planes, because burning hydrogen high up in the atmosphere creates high-altitude water vapor, which has a warming effect.

Zero-carbon cargo shipping is currently being researched, with promising results for large sails and kites. Hydrogen generated from renewable electricity

---

Biofuels: cars vs people

It sounds like such a nice idea at first. Instead of nasty polluting crude oil, why not use plant oils and sugars to power our vehicles? The crops will take in CO_2 as they grow, and release it when they're burned, thus not putting any extra carbon in the atmosphere, and we can keep on growing crops indefinitely, so the fuel won't run out. Think about it a bit more, though, and lots of obvious problems start cropping up (if you'll pardon the terrible pun). Agrofuels (the more accurate name for biofuels) require land – and the thing with land, as Mark Twain once said, is that they're not making any more of it. More specifically, there is a limited amount of good quality farmland that can be used for growing food crops – and using it to grow fuel instead has three disastrous effects.

First, the price of basic food crops goes up, with knock-on effects around the world. The food shortages and resulting riots in Bangladesh, Brazil, Egypt, the Philippines, Haiti and many other countries in 2007/08 were caused by higher prices linked, at least in part, to croplands around the world being usurped by agrofuels. Second, agrofuel companies frequently replace small, sustainable farms with giant plantations, ruining local people's food security and livelihoods, and doing long-term damage to soils and waterways. Finally – and worst of all with regard to climate change – the demand for cropland goes up sharply, putting pressure on both agrofuel companies and impoverished farmers to clear-cut the world's forests for extra land.

These ill-effects haven't stopped governments and businesses rushing ahead with agrofuels anyway, dazzled by the idea of making an easy switch to a different type of liquid fuel rather than taking on the difficult – and expensive – tasks of sorting out public transport or developing electric cars. The UK, EU, Brazil, parts of China, and the US all have laws that require a certain amount of agrofuel to be used each year. The result has been an unmitigated disaster for people and the climate. The Indonesian rainforest, in particular, is now being hacked down and burned at a terrifying rate to make way for oil palms, destroying the lands of indigenous peoples and threatening iconic species like the orang-utan.

It's a Bit Like… Realizing that the smoke from your coal-burning fireplace is polluting your neighbor's farmland, you decide to solve the problem by ripping up all of her crops and burning them in your fireplace instead. Followed by all the food in her house, some ancient woodland and a few endangered species for good measure. ■

What are the solutions?

could also be an option. Shipping emissions should also fall as we shift to an economy not based on moving loads of consumer junk around the world.

Airships might just be coming back into fashion. They use far less fuel than airplanes, and fly low enough that they could burn renewably generated hydrogen without affecting the climate. They could be a rather graceful way to cross the oceans, although rather slower than airplanes – it currently takes 43 hours to cross the Atlantic on a modern airship.

Smarter living

Better building design can make a huge difference to the energy used for cooling and heating – there are homes in Germany that can maintain a comfortable temperature with no energy input at all. Existing buildings can be hugely improved with better insulation and ventilation.

More efficient appliances, lighting and computers should become the norm.

Ground source energy isn't just for people living next to volcanoes. Even small temperature differences between the ground and the air can be exploited by ground-source or air-source heat pumps. They work a bit like a fridge in reverse, pumping in heat from just under the ground to warm a building. They use a small amount of electricity, which needs to come from our renewable supply.

Solar water heating works even in cool weather, so long as the sun is shining. The sun's rays are collected by a highly conductive black surface to heat up water.

Biomass – burning wood for heating – can work on a small scale, if there's a sustainable local wood supply. It's only really appropriate in these specific situations. An unregulated mass market in wood fuel would almost certainly lead to an increase in deforestation, which is why burning it in power stations is bad news. Similarly, so-called next generation biofuels (made from grasses and other plants that can be grown on land not suitable for crops) might be suitable for small-scale local use if carefully controlled, but a global market in the stuff would inevitably lead to mass monocultures being forced on the Global South, and people being pushed off their land.

Biogas is waste methane collected from sewage and landfill sites, or made in a digester from food waste.

Carbon capture

Wouldn't it be great if we could just make all those pesky fossil-fuel emissions disappear? If we could just suck the CO_2 out of the power stations, steel works and tar sands refineries, bury it underground somewhere, and just keep on burning coal, gas and oil into the future? Wouldn't that be a brilliant solution to all our problems?

Well, if you happen to live next to an opencast coal mine, tar sands extraction project, oil well or gas pipeline, probably not. But if we forget about all the people who are being negatively impacted by fossil-fuel extraction and just concentrate on stopping climate change, wouldn't it be a great idea?

Governments and energy companies certainly think so – they've been pouring time and money into Carbon Capture and Storage (CCS) research, and talking it up at every opportunity. Listening to pronouncements from energy and environment ministers in the UK, the US, Canada, Australia or China, you'd be forgiven for believing that this technology is just on the horizon and is about to solve all our climate change problems. Sadly, the reality isn't quite so promising.

Tried and tested carbon capture that could be safely fitted to a large power station is still a long way off (if it's ever going to happen at all). The most optimistic industry experts reckon it'll arrive in 2030; others make it 10 or 20 years later. In other words: way too late for avoiding runaway climate change. It looks like we need to stick to our original targets and keep the fossil fuels in the ground.

Despite this, governments and business keep talking about carbon capture as though it's an imminent solution to everything. Cynical people like me can't help noticing that this is a brilliant distraction tactic, allowing them to fob off public concern while continuing to burn coal and dig up the tar sands. Good old business as usual.

It's a Bit Like... You're driving towards a cliff edge, faster and faster. The people in the back seats are shouting at you to turn the car. You shout back: 'It's just not politically possible to change direction at this point! But don't worry, I'm pretty sure these untested home-made bolt-on wings will be ready in time.' ∎

What are the solutions?

It's a way of catching methane that would otherwise escape as a greenhouse gas, and putting it to use. It's a handy fuel, but a limited one – we can make much bigger emission cuts by producing less waste in the first place. Because it's possible to build a simple digester in a garden, yard or community space, it's particularly useful for providing off-grid heating and cooking fuel.

Electric heating and cooling generated from renewables can fill in the gaps wherever other, more efficient forms of heating and cooling aren't available.

Construction materials for buildings can be replaced with, or supplemented by, natural products like wood (so long as they're from sustainable sources). These products take less energy to produce, and – because buildings stay up for tens or hundreds of years – the carbon in the wood stays safely out of the atmosphere for a helpful chunk of time. Low-carbon cements have also been developed – which is fortunate, because standard cement produces *loads* of emissions.

On the make

Manufacturing can and should use a lot less energy. The main way to achieve this will be by making a lot less superfluous stuff. Do we really need plastic wrappers on bananas, electric nose-hair trimmers, 50 brands of bottled water, a new pair of jeans every week, life-sized inflatable penguins and a plastic toy free with every burger? There's plenty of evidence that accumulating more and more stuff doesn't make us any happier – once we've got the basics, we value time far more than we value possessions. It's a horrible cliché but it seems to be true – people really do prefer spending time with their friends and family, doing things and experiencing life rather than accumulating stuff.[7] In the industrialized nations, we've become trapped in a cycle of working ever longer hours to earn more money to buy more things that we don't need, while missing out on the things that actually make us happy. So our first saving in the manufacturing sector comes from making, and selling, less junk. We can also make big savings by buying and selling things second-hand, giving stuff away when we don't need it any more, repairing things and recycling more.

Most factories could also become a lot more efficient – there are plenty of examples of manufacturers slashing their energy and raw material costs by upgrading equipment and improving processes. The products themselves can be made more lightweight, durable and repairable, and include more recycled content. Producing more goods locally will cut down on transport energy and take exported emissions back from manufacturing-heavy countries like China.

Powering up

Renewable electricity is on the rise – the first six months of 2010 saw more investment in renewables than ever before. It's still nowhere near enough, but it means that there's enough of this technology out there to know that it works. Designs are improving, and prices are beginning to fall.

With all renewables, *variability* can be a problem. Unlike a fossil-fuel power station, you can't just switch renewables on whenever you need them. The wind isn't always blowing in the right place, solar panels don't work at night, a dry spell might slow down your hydro power. There are two technical ways to deal with this: storage systems and demand management. The latter is the most efficient, and includes measures like charging vehicle batteries and electric storage heaters at night when demand is low. Storage technology still needs more development, but is improving all the time. Electricity sharing between countries can also help with variability – the wind is always blowing *somewhere*. However, placing large amounts of renewable capacity in one country and using it to power neighboring countries is a different

What are the solutions?

matter, and could create the same kinds of political difficulties that we currently have with oil-rich states.

Solar photovoltaic (Solar PV) – Silicon-based cells that transform sunlight energy into electricity. They work best in sunnier countries, and are also good for providing a local power source in remote areas.

Wind power – Those big, graceful turbines that rural campaign groups love to hate. A very promising source of renewable power, but they need to be properly sited in a windy spot, and are most efficient when supersized and arranged in giant farms. Onshore turbines are easier to erect, cheaper to run and can share the space with sheep or cattle, but need to be carefully sited to cause minimum disruption to wildlife. Offshore turbines can be bigger and thus more efficient, can tap into more reliable wind supplies and are less likely to face local opposition, but have higher installation and maintenance costs.

Tidal power – The tides are driven by the moon's gravity, so this is effectively lunar power. It often involves erecting barrages or lagoons to concentrate the power of the tide onto underwater turbines. This has the potential to disrupt wildlife if not done sensitively (this is a major concern with the proposed Severn Barrage in the UK). On the plus side, tidal power is a completely predictable and very reliable source of renewable electricity.

Wave power – Waves are created by the wind, so this is essentially a form of waterborne wind power. Floating devices are moved by the waves, generating electricity. This is a pretty new technology, with only a few devices currently operating (including a 180-meter sea snake device off the coast of Scotland), but has lots of potential for rapid expansion.

Hydropower – Small-scale hydropower – turbines placed in fast-flowing rivers or streams – has a lot of potential in hilly areas. This could be particularly useful for supplying power to remote mountainous regions of the world. They need to be carefully sited to avoid disturbance to wildlife. Large scale hydropower – giant dams – is a different matter entirely. They displace communities, destroy habitats, and can create more greenhouse gases than they save via methane from

What about nuclear?

So long as we can reduce our electricity use to a sensible level then we won't need nuclear power. This is definitely good news – nuclear power is polluting (leaks and accidents still occur all over the world*), produces waste that there's no safe way of storing, requires destructive uranium mining, and makes it easier for governments to build nuclear weapons. None of that is quite as bad as climate change (unless there really is a nuclear war), but it's definitely stuff we want to avoid. Nuclear power stations are also notoriously expensive and time-consuming to build; they usually go hugely over budget and take years and years to finish. Wind, solar, tidal, wave and small-scale hydro power are far quicker to get up and running. We're on a tight timescale here, and spending our limited time and money on energy reduction and renewable power is far more likely to get us off fossil fuels in time than expensive and clunky nukes. We probably need existing nuclear stations to run to the end of their lives to give us more time to get the renewables in place, but building a load of new ones would take time, funding, and political effort that would be better spent in other ways. Meanwhile, uranium is a finite resource and will get more expensive and hard to find if the nuclear industry starts rapidly expanding.

Nuclear power is popular with many governments because, surprise surprise, there's a powerful nuclear industry lobby, and the dubious promise of nuclear power lets politicians pretend that we can keep on with good old business as usual (see Chapter 8). Supporters of nuclear power talk excitedly about the imminent arrival of fourth-generation plants, which will produce far more power, will run on spent fuel, will need less uranium and produce less toxic waste. The trouble is, as with carbon capture, we don't have time to hang around and wait for these new technologies – even the nuclear industry itself doesn't expect them to be ready for commercial construction until 2021 (for one design) and 2030 (for the rest) – and that's if everything goes to plan, which would be unusual for nuclear power. If these wonder technologies ever appear, then we can factor them into our plans and decide whether we want them – but we're in the middle of a climate emergency, and we need to get started with the solutions that we've got right now. ■

Further reading: Corporate Watch, *Broken Promises: Why the nuclear industry won't deliver*, www.corporatewatch.org.uk

* See http://nin.tl/cAaNWE

What are the solutions?

rotting plant matter. They should be avoided wherever possible.

Geothermal electricity – The power of hot rocks. Not many places have enough geothermal power to generate electricity, but it can be extremely effective on a local level – Iceland heats 87 per cent of its buildings and generates 24 per cent of its electricity using the heat from its volcanoes.

Concentrating solar power – This is a technology for spacious sunny areas (such as deserts) – it uses mirrors to concentrate light onto a central point, which then contains a small steam-powered turbine or a photovoltaic cell, which generates electricity. It takes up more space

Geoengineering and technofixes

But why fiddle around with all these different technologies and difficult social changes when you could just launch a few big technologies to sort it all out? There are various things being researched: genetically modified algal fuel, launching mirrors into space to deflect some of the sun's rays, discovering reliable nuclear fusion, dumping iron or other substances in the oceans (to increase the amount of carbon they absorb) and spraying sulphates in the sky (to mimic the cooling effect of sulphate air pollution).

These kinds of large-scale technical solutions usually promise huge expense and no guarantee of success. Even if they worked, most of these schemes would place disproportionate control of the global climate in the hands of a small number of companies or governments. Imagine if the US or Chinese government had control of a giant space mirror that was the only thing preventing the world from being fried, or if Monsanto held the patent for an algal fuel that the whole world relied upon for power. What a beautiful future we'd be building.

Some technologies, like algal fuels and nuclear fusion, are very unlikely to be available in time to help us. Others – particularly those that involve tampering with the Earth's natural systems (known as geo-engineering) – involve all sorts of unknown dangers. We still don't fully understand the complex relationship between ocean food webs, currents, temperatures and climate. Tinkering with the oceans' chemistry could have all kinds of unexpected and disastrous effects, such as poisonous algal blooms, mass species die-offs or even the faster release of marine CO_2 into the air. Spraying sulphates into the sky could have unpredictable effects on cloud formations and trigger outbursts of acid rain. Whether they'd have any significant cooling effects is unknown.

The most dangerous aspect of all these schemes, however, is that they per unit of power than normal PV panels, but each unit is much cheaper, especially with the steam turbine version. They're a good addition to our renewables mix, but they do require a lot of space as well as water, which tends not to be abundant in deserts.

Making it happen
It's going to take a lot of energy and resources to make all this sustainable technology. The Zero Carbon Britain report recommends building over 130,000 offshore wind turbines to power the UK, each up to 100 meters tall, covering an area twice the size of Wales. Zero Carbon Australia suggests that 12

give governments and businesses an excuse for inaction. Those opposed to change (see Chapter 8) can use the false promise of big shiny technical fixes to keep the fossil-fuel profits flowing for a little while longer.

The only serious argument in favor of technical fixes is a fatalistic one: what if we don't succeed in reducing greenhouse gases? If we reach a point where CO_2 feedbacks start to kick in hard and we realize we're heading for runaway warming, then mightn't we want to try some last-ditch emergency sulfur spraying or ocean tinkering? In which case, shouldn't we be researching them now so we can use them as safely as possible if we need to? Well, maybe – but first we need to make sure we're doing all we can not to reach that theoretical future point. The risk is that what starts out as just-in-case research becomes used as an excuse for not cutting CO_2, or takes people and funds away from working on the real solution: keeping fossil fuels in the ground. The world would somehow need to agree a way to keep tight control over the technology, to prevent it from being launched too early by a desperate government or unscrupulous company. There's also no spare Earth for us to practice on – we can never be sure what effect a large-scale atmosphere or ocean project would have unless we try it, and then it's too late.

Geoengineering ideas are gaining popularity in some quarters – there have been pro-sulfate-spraying demonstrations in Australia, and the growing market in carbon emissions could help to boost some of these schemes (see Chapter 9).

It's A Bit Like... Your house is on fire, so you sit down in the living room and start drawing up designs for a giant wall-smashing robot. ∎

Further reading: Corporate Watch, 'Techno-Fixes' report, www.corporatewatch.org.uk

What are the solutions?

concentrating solar thermal plants will be needed for the country, with a total area of 2,760 km², about the size of Kangaroo Island. This needs to happen all over the world – not always in such a centralized way, but still requiring huge amounts of materials and energy in total. This will have unavoidable negative environmental impacts which we'll need to minimize as best we can. We'll have to place everything as efficiently as possible, make no more than we need, use as much recycled material as we can, and rig up our wind turbine and solar panel factories to run off renewable energy as soon as possible.

Food

According to a ground-breaking report from GRAIN that pulls together the work of many different scientists, the food industry is responsible for somewhere between a third and a half of global emissions.[8] This includes the manufacturing of chemicals and fertilizer, carbon loss from soils, farm energy use, transport, processing, refrigeration, and the clearing of forests to create agricultural land. All of these effects are heavily associated with large-scale industrial farming and not with small-scale organic or peasant farming.

In fact, a global shift back towards traditional farming practices could reverse the loss of soil carbon caused by chemical fertilizers, and start pulling CO_2 from the atmosphere back into the soils. GRAIN's conservative estimate is that traditional management of the world's agricultural land could draw 3,750 million tonnes of CO_2 per year out of the atmosphere over the next 10 years – equivalent to almost 10 per cent of current global emissions. Combined with the reduction in fossil-fuel use from cutting out the chemical inputs, this could give a huge boost to our chances of getting down to 350 ppm of atmospheric CO_2. The precise scale of these savings is still being debated by soil scientists (the numbers above are just best-guess estimates), so we shouldn't take them into account in our carbon targets; they're an added extra that could nonetheless be the difference between success and failure.

But don't we need high-input farming to feed the world? Fortunately, the answer is no – a better-organized and less wasteful global food system could feed us all using traditional and organic methods.[9] Industrial agriculture has boomed not because it is necessary but because the low price of oil and lots of government subsidies have made it highly profitable in the short term. Industrial farming is, in fact, hugely inefficient – according to the UN Food and Agriculture Organization, farmers in industrialized nations use five times more energy to produce a kilo of grain than farmers in Africa. This doesn't even include all the extra energy for transport, processing and freezing. Chemical fertilizers strip organic matter from the soil, reducing its long-term fertility and making it more vulnerable to erosion and flooding. In addition, the industrial food system throws away fully half of all the food it produces. This is enough to feed the world's hungry six times over, every year.

Traditional or organic farming uses far less energy but significantly more labor. If a farm is producing food to sell in the global marketplace, this looks like a problem because it increases the costs of production and makes it harder to sell the produce at a profit. However, if farmers are producing food for themselves and for their local community then it's not a problem at all – it increases employment, gives more people a share in the produce and provides independence from volatile global food prices. Sustainable agriculture is therefore more likely to succeed when practiced on a small scale, with the aim of feeding people rather than making a profit. This has the added benefits of reducing global hunger by giving people control over their own food supply (most hunger is caused

What are the solutions?

by unequal access to food, not food shortages), and massively reducing the need for global food transport, processing and refrigeration.

There is no simple one-size-fits-all solution for food. Different practices work better in different countries and regions. The best solution in almost every case is to allow small-scale local farmers to use traditional/organic, non-chemical practices to produce food and restore soils at the same time, supported by whatever labor-saving renewable energy technologies are appropriate to their needs. Most food would be for local consumption – we'll still want some trade to keep our diets varied, healthy and interesting, but on a much smaller scale than today. Southern nations would primarily grow food to feed their own populations, rather than cash crops for export.

Intensive livestock production is inefficient, energy-intensive, produces large amounts of methane, and is a key cause of deforestation – not to mention incredibly cruel to the animals involved.[10] We'll need to transform livestock farming from a huge, voracious global industry into something smaller-scale, less intensive and properly integrated with other farming practices.*

This approach is supported by millions of farmers around the world, many of whom are part of the powerful Via Campesina peasant farmers' network. It would reduce global hunger, greatly reduce emissions and also have huge benefits for our health here in the rich world – a diet based on fresh fruit and vegetables with only a small amount of meat and dairy is recommended by Northern doctors for a long and healthy life.[11] Most of the world's farmers are

---

* As a vegan myself, I have other ethical concerns about using animals for food (as do many people), but this isn't the place to explore them. Looking at this purely from a climate change perspective, we don't need to cut out meat and dairy entirely, just eat a lot less of it, and produce it in a very different way. For more about veganism, please do see www.vegansociety.com

women, and supporting their struggles for lands and livelihoods is also a vital part of the battle for women's rights internationally.

Protecting forests

When it comes to forest preservation, who better to ask than the people who've lived in the world's forests for thousands of years?

Indigenous peoples – in North and South America, Africa, Asia, and the Pacific – have developed the art

---

Won't peak oil save us?

Oil is the fossil fuel that's in the shortest supply. New finds are becoming rarer and rarer, and when they are discovered they're usually smaller than previous finds. More and more energy analysts are talking about Peak Oil – the point where the annual demand for oil becomes higher than the supply, causing prices to rocket. Some say it's 20 years away, and some say it's already arrived. One thing's for sure, though – all the easy oil has already been found, and the general trend in oil prices is going to be upwards from now on.

This is good and bad news for the climate. On the plus side, rising oil prices help to lower demand, and so non-essential oil use drops. People drive a bit less, fly a bit less, and transport goods around a bit less. However, industrial societies are still extremely reliant on oil to function – especially our transport and food production. Until we make some serious changes to our infrastructure, oil demand will still remain high, and so those higher prices have made expensive, dirty, CO_2-intensive and risky oil sources like tar sands, oil shale, coal-to-oil and deep offshore drilling more financially viable. Higher oil prices are likely to make climate change worse in the short term – and the short term is all we have.

The other problem is that our global renewables rollout is going to need lots of transport energy, and in at least the early stages that will have to come from oil. The higher the price of oil, the more expensive the transition to a zero-carbon world will be. Meanwhile, the faster we can shift to non-oil-based transport and food production, the less hard we'll be hit by peak oil when it does arrive. So peak oil, rather than being a climate solution, gives us even more incentive to get off the fossil fuels as soon as possible.

It's a Bit Like... Hoping that peak oil will save us from climate change is like hoping that the sharks banging on the hull of our leaky, slowly sinking boat are here to save us from drowning. ∎

What are the solutions?

of sustainable forest living over many generations. With only a few exceptions, mass deforestation has been driven by governments and corporations seeking minerals, fossil fuels, and timber. Around the world right now, indigenous peoples are defending their forests using every means they can – from courtroom battles to direct action. Research has shown that recognizing indigenous forest people's land rights would be a highly effective and inexpensive way of protecting the world's forests.[12]

The fairer food system described above, based on feeding people rather than making profit, would also be a huge step towards protecting forests; desperate landless farmers would no longer be chopping down trees to create space for farmland.

Reducing our consumption in the North would reduce the demand for cheap minerals and timber that drives so much deforestation. We'll also need Northern nations to make up for their historical climate change responsibility by paying to preserve and restore the forests of the South – and do quite a lot of replanting in their own countries, too.

A better world is possible

It is possible to have a zero-carbon world, and if we do it properly there should be loads of other benefits too: safe, comfortable homes with lower energy bills, better quality food, healthier populations, a massive reduction in air pollution and respiratory disease, safer streets, and cheap and reliable public transport. These benefits should be made available to everyone, not just the wealthy few. Millions of small-scale farmers and indigenous people would also have their lands and livelihoods secured, leading to a significant decrease in poverty, hunger and injustice all over the world. Achieving all this won't be easy or painless – but it's definitely possible.

To figure out how to get to this zero-carbon future,

we first need to have a look at how we got into this mess in the first place...

1 David Mackay, *Sustainable Energy – Without the Hot Air*, UIT, 2009, www.withouthotair.com; PriceWaterhouseCooper, *100% renewable electricity: a roadmap to 2050 for Europe and North Africa* 2010, available at www.pwc.co.uk **2** Their *Zero Carbon Britain* report (www.zcb2030.org) gives a total of 13,000 KWh/year. However, Britain has exported a lot of its manufacturing, so I have upped this figure slightly to take account of that, to 15,000 KWh – an increase in line with a similar estimate made by energy expert David Mackay (see footnote 1). **3** See David Mackay, footnote 1. **4** These figures don't include energy from biomass. **5** R Dahl, 'Vehicular Manslaughter', *Environmental Health Perspectives*, 2004, http://nin.tl/bSx6Hb **6** UK Civil Aviation Authority passenger surveys. **7** www.newdream.org/about/poll.php **8** GRAIN (2009), http://nin.tl/ciRh9q **9** Colin Tudge, *Feeding People is Easy*, Pari Publishing, 2007. **10** Pelletier and Tyedmers, 'Forecasting potential global environmental costs of livestock production 2000–2050', *Proc Nat Acad Sci*, 107, 2010. **11** See, for example, http://nin.tl/9eU8TN **12** Accra Caucus, *Realizing rights, protecting forests*, http://nin.tl/9PV86b

Part C: The Way Forward

Lifestyle Choice

Feeling hardy in my cardy
Armed with insulation foam
I pit my wits against the cracks and splits
That fill my leaky home.
With a mighty laugh I block the draught
Beneath the kitchen door:
Just another night of fighting
In this endless eco-war.

I boldly scout the shopping out;
I am the kind of buyer
Who researches every purchase
For the dirt on the supplier.
I cycle, I recycle,
I have more faith than George Michael
That getting my own lifestyle right's
The only way to win this fight.
Even though my friends and neighbors
– Strangely unmoved by my labors –
Keep taking reckless impulse flights
Leave their TVs on all night
And defend their god-given right
To sixty thousand Christmas lights.

Until one day's moral tussle
– Local or organic mussels? –
Pushed guilt fatigue to new extremes:
That night, I had the strangest dreams.

I dreamt that Martin Luther King
Was standing by my compost bin
And with fine, impassioned words
Told me which methods he preferred
To keep the fruit flies out, and the kitchen peelings in.

But while he gave my crazy paving
Speech after noble speech
On his dreams of different compost schemes
(With pros and cons for each)
He wasn't firing up Americans
On justice, class and race
And without his flame the world became
A sadder, darker place.

That vision burst; I fell headfirst
Into another scene
Where Chris and Sylvia Pankhurst,
Their mother Emmeline,
And a thousand other suffragettes
Sought recognition and respect;
But not with shouts and chants and chains
And hunger strikes to stake their claims:
They sought the same effect
By being sure, when at the store, to only ever choose
Kitchen products guaranteed to be
Empowering to use.

Then I saw Gandhi getting handy with a pair of
bathroom pliers
To defend his independence from the British occupiers
Not by mobilizing peasants into peaceful mass resistance
But by fixing leaky taps and putting bricks inside their
cisterns.

While the abolitionists, instead of fighting slavery,
Just stayed at home and put a bit less sugar in their tea.

And Che Guevara
Avoided the palaver
Over 'was he a hero or a merciless killer?'
By staying indoors
Down on all fours
Launching attacks on the skirting cracks with
eco-polyfilla.

I woke up with a start, with pounding heart, my body aching
And my belief in the *Observer*'s lifestyle pullouts badly shaken.
Even I could see
That eco-piety
Is great for scaring off speed dates but it won't change society.

I should be the change I want to see
Try to avoid hypocrisy
But taking one car off the road won't make more trains and buses run
Stopping up one leaky home leaves countless millions to be done.
If I want to be noticed
By the 'people in power'
Should I join a mass protest
Or fit a new shower?
And getting in a panic
If my balsamic's not organic
Is less useful and more draining than just *doing some campaigning*.

So,

Now I've met people all around me I might otherwise have missed
Young, old, often well-respected unexpected activists.
They don't just care – they are prepared to take the action that's required
My plumber and my mum are feeling equally inspired,
We'll fight the problems at the top, and build solutions from below
With humor, hope and energy, and this is how I know
That if you want to stop the tales of climate doom coming to pass
Yes, reduce your carbon footprint – then use it to kick some ass.

7 What is the political history of climate change?

The beginnings... The first signs of danger... Climate politics hots up... Kyoto, Copenhagen, carbon trading and hot air.

1400-1800: Early history

1400s to 1600s – Across Europe, common land is enclosed and put into private hands. Peasants are forced to work as laborers, producing high-value crops for the wealthy instead of farming to feed themselves. There is a general shift of power away from the monarchy and the church and towards increasingly wealthy merchants and entrepreneurs. European settlers begin to colonize other lands, driven by the search for new fertile soils – and people – to exploit for profit.

1700 to 1800 – The 'Enlightenment' in Europe sees the birth of modern science, but also a general shift in attitude towards nature. Rather than seeing humans as part of nature, Europeans increasingly see nature as something to be conquered and exploited. Powerful Europeans see themselves as 'superior' to the rest of the world's people, and use their new-found technology to brutally occupy much of the Americas, Africa and Asia. The colonies – and the peasants and workers of Europe – are treated as a source of labor, fuel and raw materials to support the needs of a minority of wealthy Europeans. Things are not going well.

This is how what we think of as the 'normal', 'Western' economic system comes to dominate the planet. These early years set the scene for what is to follow, by putting in place 'a new global system based on industry, consumption, and an endless quest for material wealth'.[1]

1800-1960: Warning signs

1800 – Carbon dioxide stands at 290 ppm in the atmosphere.

1824 – Joseph Fourier calculates that the world would be far colder without an atmosphere (see Chapter 1).

1850s – The Industrial Revolution gets properly under way. Greenhouse gas levels are on the rise.

1859 – John Tyndall figures out the heat-trapping properties of greenhouse gases (see Chapter 1).

1896 – Svante Arrhenius links humanity's carbon dioxide emissions to changes in the world's climate (see Chapter 1).

1897 – US geologist Thomas Chrowder Chamberlin works out a model of global carbon flows.

Early 1900s – Electricity, chemicals, and public health all come on the scene, increasing greenhouse gas and population levels. US industry is booming, fed by the huge natural wealth of the continent, and the US's role as a major global power is now well established.

1917 – The Russian Revolution leads to the creation of the Soviet Union. Between now and its final collapse in 1991, the Soviet Union will emit 110 billion tonnes of CO_2 – about 10 per cent of all the carbon dioxide ever released by humanity – thanks to its leaders' relentless focus on highly polluting industrial growth.

1920 to 1925 – The first Texan and Persian oil fields open. The era of 'cheap energy' has arrived.

1938 – English engineer Guy Stewart Callendar notes that the world has been warming and suggests it could be due to rising CO_2 levels. His work gets into a few school textbooks but no-one pays much attention.

1939 – The *Time* magazine cover story for 2 January is called 'This Warming World'.

Mid-1900s – Most former colonies in Asia, Africa and Latin America have now gained nominal independence, but land and resources in the Global South are still largely under the control of wealthy minorities, working together with Northern companies and governments. Fossil fuels, agricultural products, timber and minerals mainly flow from South to North, in unfair trade relationships that provide cheap resources to corporations and consumers but keep Southern workers and farmers in poverty. The average income in the North is now 20 times the average income in the South. Things still aren't going well.

1950 – CO_2 emissions stand at 5,800 million tonnes per year. Around 90 per cent of these emissions are from Europe, Australia, Russia, Canada and the US (with just 28 per cent of the world's population).

1950s – Happiness in wealthy Northern countries reaches its highest level.[2] From this point onward, people in industrialized nations consume more and more, but don't (on average) become any happier.

1956 – Physicist Gilbert N Plass calculates that humanity's CO_2 emissions are on course to raise the planet's temperature by 1.1°C per century (a prediction that wasn't too far off the mark). The *New York Times* 'Science in Review' section reports that 'Warmer climate on the earth may be down to more carbon dioxide in the air'.

1958 – The US Weather Bureau starts using Charles Keeling's new technique for measuring atmospheric CO_2 at Mauna Loa. A popular documentary called *The Unchained Goddess*, produced by Frank Capra and shown in schools all over the US, predicts that humanity's CO_2 will lead to the melting of polar ice and rising sea levels.

1961-1986: Warming up

1961 – Carbon dioxide in the atmosphere is 318 ppm, and annual CO_2 emissions have reached 9,200 million tonnes per year. Around 82 per cent of these emissions are from Europe, Australia, Russia, Canada and the US.

1965 – US President Lyndon Johnson's Science Advisory Council warns that increasing CO_2 'will modify the heat balance of the atmosphere to such an extent that marked changes in climate could occur'.

1960s and early 1970s – Despite occasional warnings from certain scientists, CO_2-induced warming is generally seen as only a minor concern. Based on the limited data available, the overall warming trend seems to have reached a plateau (thanks to temporary 'global dimming' caused by sulfate pollution – see Chapter 1). Meanwhile, thanks largely to the continued flow of cheap raw materials from Southern nations, transnational corporations grow in wealth, power and influence. New hybrid crops and mass-produced chemical fertilizers are rolled out across the world in a so-called 'Green Revolution' that increases short-term yields and boosts the profits of agricultural corporations, but greatly increases the greenhouse gas emissions of agriculture and reduces the long-term fertility of soils.

1973 and 1979 – Oil-producing states restrict production, and the resulting 'oil shocks' cause energy prices to rise steeply. In Europe and North America, this triggers public action on energy conservation and a demand for alternative energy sources. This helps to strengthen a growing 'environmental movement' in the industrialized world. Meanwhile, oil-producing states receive huge amounts of money from these higher prices, much of which they place in Northern banks. The banks, flush with cash, start providing massive loans to Southern governments, often to dictators and corrupt regimes. This launches the so-called 'Third World Debt Crisis', saddling many Southern nations with crippling interest payments for decades to come. This puts them under even greater pressure to raise cash through fossil-fuel extraction, intensive agriculture and deforestation, whatever the cost to their people, local environments, and the climate.

1974 – Carbon dioxide in the atmosphere hits 330 ppm. The average income in a Northern country is now 40 times the income of someone in the Global South.

Late 1970s to 1980s – Rising CO_2 levels outweigh the cooling effects of sulfate pollution and global temperatures begin to climb once more. Research into ice core samples helps to unlock the secrets of prehistoric warming cycles, and the link between CO_2 and huge climatic changes. The role of other greenhouse gases such as methane is revealed. Climate science leaps forward in sophistication, and scientists start to take the risks of human-induced climate change very seriously indeed.

1981 – President Reagan is elected in the US as part of a political backlash against the environmental movement. A link is formed between political conservatism and global warming denial that has never really gone away. The scene is set for numerous clashes between scientific bodies and the US government over the seriousness of climate change throughout the 1980s.

1983 – Climate change is mentioned for the first time on the UK national evening news.

1980s to 1990s – Corporations take advantage of cheap Southern labor by relocating factories and sweatshops to countries where lax regulations and crippled local economies make workers easier to exploit. Encouraged by US and European governments with strong ties to polluting industries, Northern nations undergo a 'deindustrialization' process which shifts a large chunk of their carbon emissions (and other pollution) to the Global South. International financial institutions like the International Monetary Fund (IMF) force Southern governments to accept this process (see Chapter 8).

1986 – Global annual CO_2 emissions reach 20,000 million tonnes. Some 67 per cent of these emissions are from Europe, Australia, Russia, Canada and the US.

1987-1997: Climate politics hots up
1987 – CO_2 concentrations in the atmosphere reach 350 ppm.

What is the political history of climate change?

1988 – NASA scientist James Hansen testifies before the Energy and Natural Resources Committee of the US Senate. He says: 'The global warming now is large enough that we can ascribe with a high degree of confidence a cause-and-effect relationship to the greenhouse effect... The greenhouse effect has been detected, and it is changing our climate now.' He also presents projections of future warming and changes in weather patterns (which have subsequently proved to be pretty accurate). Combined with record high temperatures and droughts, this sparks a leap in media coverage of the issue. A major climate conference in Toronto leads to the founding of the IPCC, to assess 'the scientific, technical and socioeconomic information relevant for the understanding of the risk of human-induced climate change'.

Late 1980s – British Prime Minister Margaret Thatcher makes a series of speeches on the threat of climate change. Cynics note that the topic gives her a convenient weapon in her political battle against the UK's coal mining unions. Her Conservative government goes on to announce one of the biggest road-building plans in UK history.

1990s – Climate science progresses rapidly, including in the areas of prehistoric temperature changes and the importance of feedback loops. Data collection improves, and scientists' understanding of the role of plant and soil carbon gradually increases. Climate change denial rises in response, often via fossil-fuel funded thinktanks and foundations.

1990 – The first IPCC report is released.

1991 – Protesters block road-building work at Twyford Down in the UK, launching a major anti-roads campaign that spreads to sites across the country.

1992 – At the UN Earth Summit in Rio, 190 nations sign up to a process called The Framework Convention on Climate Change (UNFCCC). It establishes a set of principles for future action on climate change, and countries agree to measure and monitor their emissions. They don't agree to any binding reductions, though, thanks largely to obstruction from US negotiators and fossil-fuel lobbyists.

1993 – President Bill Clinton and Vice-President Al Gore take office in the US. Clinton pledges to reduce US greenhouse gas emissions back to 1990 levels by the year 2000. An energy tax is proposed, but dropped after pressure from industry lobbyists and representatives from oil- and coal-producing states. Other potentially climate-friendly reforms are similarly retracted or watered down, despite public support for many of these measures. By the end of Clinton's presidency, US annual CO_2 emissions have risen by 18 per cent.

1995 – The second IPCC report is released, and the first UNFCCC 'Conference of Parties' takes place in Berlin. Small island states and Southern nations push for binding emission reduction targets, and manage to win support from the EU. The US government finally agrees to reduction targets, but also starts lobbying for an emissions-trading scheme that will allow rich nations to 'buy' emissions savings from poorer nations. CO_2 levels reach 360 ppm. Ken Saro-Wiwa – a passionate campaigner against Shell's drilling activities in the Niger Delta – is executed by the Nigerian government on trumped-up charges, sparking international outrage.

1996 – Following five years of concerted environmental protests, the UK government withdraws most of its proposed road-building plans.

1997 – The Kyoto Protocol is drawn up through the UNFCCC process, requiring the richest nations to start making cuts in their greenhouse gas emissions (see box). Under huge domestic pressure from fossil-fuel lobbyists (including a $23-million anti-Kyoto television campaign) the US delegation, led by Al Gore, only signs up on the condition that emissions trading is included in the deal.

What is the political history of climate change?

1998-2009: Hot air

1999 – The World Trade Organization (WTO) meets in Seattle to try to expand its powers, which already include the ability to override environmental and social regulations in the name of corporate profits (see Chapter 8). Huge street protests from a diverse coalition of Northern and Southern activists lead to the collapse of the trade talks. Further WTO meetings throughout the 2000s are subject to similar protests, and the talks fail to pick up again.

2000s – Climate change protest spreads across the world. Direct action against the coal industry becomes commonplace in Australia, Canada, Germany, New Zealand/Aotearoa, Britain and the US. Farmers block

The Kyoto Protocol

Forging a global agreement to tackle climate change was never going to be an easy task. As shown by the not-terribly-cheerful timeline on these pages, the whole thing's made hideously complicated by massive global inequalities, differing levels of responsibility, and a history of exploitation and broken international promises.

Perhaps, then, we shouldn't be too surprised that the international climate negotiations have not yet achieved a workable global solution. The best they've managed so far has been the 1997 Kyoto Protocol, under which industrialized nations (known as 'Annex 1' countries) pledged to cut their CO_2 emissions by a completely inadequate 5.2 per cent below 1990 levels by 2012. The US famously pulled out of the deal, and most of those who remained in are unlikely to achieve even these small cuts.

Meanwhile, no definite plan has been agreed for ensuring that the poorer nations switch to a climate-friendly development path. The US says it won't play unless, in the name of 'fairness', all non-Annex 1 countries also take on emissions reduction targets. Southern governments, however, point out that they've arrived late to the fossil-fuel feeding frenzy: the industrialized nations got us into this mess by emitting, over the past 200 years, the vast majority of the greenhouse gases currently warming up the atmosphere. How can the Annex 1 countries demand that the South restrict its development with tough carbon targets when the North has mostly missed its own Kyoto goals?

With Kyoto limping to the end of its life, governments are scrabbling to strike a new deal on global emissions cuts between 2012 and 2020. This was meant to happen at the 15th Conference of Parties (COP-15) at Copenhagen, in December 2009. Efforts were focused on persuading the gas pipelines and coal plants in India; indigenous peoples defend their lands against logging companies in South America and tar sands extraction in Canada; activists unite against gas flaring in Nigeria; communities take a stand against biofuel plantations in Brazil and palm-oil deforestation in Indonesia. People involved in these 'frontline' struggles increasingly refer to the threat of climate change as part of their activism.

2001 – The IPCC releases its third Assessment Report. The new US government, led by George W Bush, pulls out of the Kyoto Protocol.

2003 – Annual global emissions reach 25,000 million tonnes of CO_2. Some 52 per cent of these emissions are from Europe, Australia, Russia, Canada and the US.

US – responsible for 30 per cent of current emissions – to sign up. But trying to forge a deal that favored the interests of wealthy nations over the real needs of the world's people was always going to be a non-starter. Any deal without a strong justice element that recognized historical responsibility was never going to be accepted by many Southern governments.

Poorer nations have fought bitterly to enshrine a 'right to development' and an acknowledgement of countries' 'common but differentiated responsibilities' within Kyoto, which means that richer countries are expected to act first. Unless the Annex 1 countries start showing real commitment to these principles – through deep domestic emissions cuts, strings-free funding, technology transfer and development allowances – the chances of the South staying on board with a post-2012 deal are bound to be slim.

Unfortunately, the trend has so far been in the opposite direction. As the climate talks progressed from Toronto (1988) to Kyoto (1997) to Bali (2007), the rich countries' targets gradually weakened by around 1,900 million tonnes of CO_2, while the role of carbon trading grew steadily.

The infamous failure of the Copenhagen talks in 2009 can be linked to these unresolved justice issues. The industrialized nations' refusal to make any significant concessions on carbon targets or global climate funds, and many Southern nations' unwillingness to sign up to a deal without these vital elements, meant that no meaningful agreement could be reached. The Annex 1 countries were joined in their obstructionism by new polluters like China in putting short-term self-interest before the need for a fair and effective global climate deal. The failure of the Copenhagen talks was ultimately down to their failure to address the key issue of climate justice. ∎

What is the political history of climate change?

Much of this shift away from the wealthiest nations is due to the export of their manufacturing to poorer countries.

2004 – Russia ratifies the Kyoto Protocol, bringing it into force.

2005 – Hurricane Katrina hits New Orleans, creating major destruction and loss of life. Concern about climate change in the US rises rapidly. The G8 leaders meet in Scotland, and name climate change as one of their top global concerns. The EU Emissions Trading Scheme is launched (see Chapter 9).

Carbon trading

This scheme allows polluting industries and governments to buy their way out of their carbon commitments, using complex trading rules written by Northern economists. It was cooked up by US negotiators back in the early days of the UNFCCC and forced into the Kyoto deal to make it more palatable for their corporate allies. You can tell this isn't going anywhere good, can't you?

The concept runs as follows: all the current annual emissions from the countries or companies in the scheme are added up – let's say it's a million tonnes of CO_2. A target is selected – say a five-per-cent reduction. This means we want to get down to 950,000 tonnes next year, so we issue that many 'emissions permits' to the people in the scheme. They can then trade those permits between each other, but they need to have enough permits to cover their pollution at the end of the year. The following year, the number of permits available – the 'cap' – is lowered again, and so on. This system is also known as 'cap-and-trade'.

It sounds fine at first until you start thinking about it properly. How do we decide how many permits to give out at the start? How do we decide on the reduction rate? Who oversees the trading and makes sure it's fair? How can we measure the emissions savings? If just a few companies make big reductions and everyone else just buys their permits and makes no changes of their own, how is that moving us towards a new, sustainable way of doing things? Why on Earth set up such a complex system when you could just pass a law that makes companies reduce their emissions?

A system like this has been running for five years in the EU, as the EU-Emissions Trading Scheme (EU-ETS). Sure enough, all the big companies involved in the scheme lobbied hard at the start and were given hugely generous amounts of permits. The rules are worked out between government officials, economists and industry representatives, none of

2006 – Global inequality widens still further: the richest 1 per cent of the world's people now have an income equal to the bottom 57 per cent. The UK sees its biggest climate march to date, with 25,000 people gathered in London's Trafalgar Square for the Campaign Against Climate Change/iCount rally. The first Camp for Climate Action takes place at Drax coal power station in northern England.

2007 – The IPCC releases its fourth Assessment Report, and Arctic sea ice reaches its lowest-ever summer extent. Public interest in climate change

whom have any real incentive to make the scheme effective. As a result, the system is riddled with loopholes – companies can secure permits from dubious 'offsetting' schemes in the Global South (see Chapter 9), and traders have made a fortune by speculating on the price of carbon. Permit prices have risen and crashed, allowing polluting companies to make large windfall profits at some times, and to collect all the permits they need very cheaply at others. When emissions fell anyway as a result of the recession, it left loads of spare permits sloshing around.

This is why, when polluting companies realized that some sort of carbon reduction law was on the way, they lobbied hard for a trading scheme. They knew it was a system they could work to their advantage. Governments were happy enough to go along with it rather than stand up to the corporations. Maybe some of them genuinely believed it would work. If so, they were wrong.

Over five years, the companies in the EU-ETS have reduced their emissions by a grand total of one third of a per cent. It's been an absolute failure as a climate strategy, but a resounding success for the profits of the companies involved – they've been able to carry on burning fossil fuels while making only minimal changes to their practices. Yet this ineffective scheme is the very same cap-and-trade that the governments of Australia and the US tried (and failed) to put in place in their countries in 2010. In these cases, it seems that the industry lobby was powerful enough to avoid even a toothless scheme like this. Meanwhile, endless discussion about carbon trading has pushed more effective solutions off the agenda of the international climate talks.

It's a Bit Like... Handing control of the Earth's vital natural systems over to a bunch of grinning Wall Street traders. Oh no, wait: that's exactly what it has done. ■

For more information: www.carbontradewatch.org, www.thecornerhouse.org.uk

continues to grow, accompanied by an explosion in 'green' consumer products. The Climate Camp at Heathrow Airport in London sparks global media interest in aviation's role in climate change. Meanwhile, the collapse of international markets, triggered by unregulated financial speculation, slides many countries into recession.

2008 – Following a major campaign by Friends of the Earth and others, the UK becomes the first country in the world to set legally binding emissions reduction targets, pledging to cut CO_2 emissions to 80 per cent of 1990 levels by 2050.

2009 – There are high expectations at the 15th major UNFCCC meeting (COP-15) in Copenhagen, fueled by increased levels of public concern, visible signs of climate change, high-profile NGO campaigns and the replacement of notoriously polluter-friendly US President George W Bush with Barack Obama. Around 50,000 people gather for 'The Wave' in London, the best-attended climate change demonstration so far – only to be overtaken later that month when 100,000 protest on the streets of Copenhagen. Direct action Climate Camps take place in around 20 different countries. However, the world's governments fail to create a binding climate treaty to replace Kyoto, which is due to expire in 2012.

2010 – The failure at Copenhagen leads to a collapse in mainstream media enthusiasm for climate change reporting, and a loss of direction for many large environmental NGOs. Climate change denial becomes fashionable again, and public concern in Northern nations declines slightly. Few governments are on track to meet even the low emissions targets demanded by Kyoto, a major climate bill is stalled in the US Senate, and the EU Emissions Trading Scheme is found to have reduced emissions by just one third of a per cent.[3] Much of the energy seems to have been sucked out of the official international process. However, grassroots activism continues to grow, and an alternative World Climate Conference takes place in Cochabamba, hosted by the Bolivian government (see Chapter 8). A huge oil spill at Deepwater Horizon off the US coast focuses renewed criticism on the oil industry, and almost 20 countries experience their hottest summer on record. Atmospheric CO_2 stands at 388 ppm, but annual global fossil fuel emissions have leveled off at around 30,000 million tonnes thanks to the recession. There's still everything to play for.

1 Patrick Hossay, *Unsustainable*, Zed, 2006. **2** National Opinion Research Center at the University of Chicago; Pew Research Center. **3** 'EU emissions trading scheme on course to make tiny savings, says report' *The Guardian*, 10 Sep 2010.

8 Why haven't we fixed it yet?

The political roadblocks that have stopped us dealing with climate change... Evaluating the various contributions of governments, corporations, international institutions, economists, campaigners, frontline communities and the media... And drawing out the common threads that can help guide us now.

HUMAN-MADE CLIMATE change isn't exactly a new problem. We've known about it for at least 50 years, and have known that it's serious since the 1980s. But throughout this whole period, greenhouse gas emissions have continued to climb – in fact, they've accelerated. All the significant dips and blips have been due to oil price fluctuations and economic downturns, not concerted action to reduce pollution.

But why? If we've known for decades that we're driving headlong towards a climate cliff edge, why haven't we (as a society) been stamping on the brakes?

To answer this question, it might help to make a quick swoop through some of the major players in this ongoing climate drama – politicians, industry, campaigners, economists, the media – and briefly consider what they've done so far to help or hinder the cause of climate safety. We'll look at the key barriers and drivers affecting their behavior, and then at the end of the chapter we'll pull together the common threads.

All the following categories are fairly general, and there will obviously be exceptions and overlap between the groups, but this approach should still be useful in helping us untangle the big picture.

Governments of the Global North

What have they contributed so far?
Well, it's been a mixed bag, obviously. But, as Chapter 7 has shown, the general trend has been for Northern

governments to range from not-very-effective to actively hostile on the issue of climate change – even though (or perhaps because) they are representing the countries most responsible for creating the climate crisis. Throughout the 2000s, elected officials have increasingly talked about the seriousness of climate change, knowing that it is of concern to the majority of their electorates, but haven't backed it up with much action. The world's governments have failed to put their own short-term interests aside and agree to a meaningful global climate treaty. As if that failure wasn't bad enough, many politicians around the world – mostly from right-wing parties – are outspoken in their climate change denial. While there are few outright deniers in government, they can be very influential in closely balanced parliaments, such as in the US, the UK and Australia.

Many government policies are actively making the problem worse. In 2008, governments admitted paying over $550 billion in subsidies to the fossil fuel industries.[1] This total does not include hidden subsidies such as tax breaks, regulatory loopholes and the huge sums spent on military deployment to defend oil supplies. By contrast, in 2009, governments spent only $45 billion on renewable energy – and that total included biofuels.[2] This is an old story – governments have channeled trillions of dollars of taxpayers' money into oil, gas, and coal companies over the years, as well as businesses involved in logging, car manufacture, airport expansion and intensive agriculture. Between 2002 and 2008, the US government alone spent $72.5 billion on fossil fuel subsidies.[3] The 2008 US Farm Bill put $78 million towards sustainable farming, but $300 *billion* – about 3,800 times as much – towards business-as-usual industrial agriculture. In 2005-06, the Australian government spent around US$9 billion on fossil fuel production and consumption, including $1.5 billion on coal.[4] By contrast, only $0.3 billion was spent on renewables and energy efficiency. All of this serves to lower the price of fossil fuels and make renewables seem more expensive.

Even supposedly environmentally friendly governments aren't doing that much better. US President Barack Obama said in 2009 that 'Climate change is serious, it is urgent, and it is growing... if we fail to meet it – boldly, swiftly, and together – we risk consigning future generations to an irreversible catastrophe.' However, after two years in power, his government has done little to meet that challenge. His administration has announced a slight increase in vehicle efficiency and a voluntary CO_2 reduction target of 17 per cent below 2005 levels by 2020. This is based on no credible scientific research and, as we saw in Chapter 5, is wholly inadequate (it's a much smaller cut than called for by even Kyoto's feeble targets).

The only other piece of US climate change legislation with any political support has been a 2010 carbon trading bill full of get-out clauses and loopholes for industries to avoid making real emissions cuts. Even this weak proposal failed to get through the Senate, despite Obama's ruling Democratic Party holding the majority of seats. Only five per cent of the $792 billion spent on reviving the US economy in 2009 was allocated to renewables and energy efficiency, with a further two per cent towards public transport. About four times as much was spent that year on roads, airports and bailing out car companies.

Meanwhile, the UK's 1997-2010 Labour government – despite passing legislation committing the country to an 80-per-cent carbon cut by 2050 – subsidized the aviation industry to the tune of $15 billion per year in tax breaks,[5] opened up new waters for deep-sea oil drilling, announced major runway and road-building plans, approved dozens of new coal mines and refused to rule out new coal-burning power plants.

Northern governments have been active beyond

their borders too, underwriting their companies' fossil fuel extraction projects to the tune of billions of dollars, through Export Credit Agencies (ECAs). These secretive government departments supply free insurance to Northern corporations digging up coal, gas and oil around the world – if these polluting projects fail to turn a profit, we, the taxpayers, cover their losses.

Despite this overall gloomy trend, there are examples of positive government action out there, especially at a more local level. California has put energy efficiency measures in place that have held their residents' per capita energy use roughly constant, while consumption in the rest of the US has almost doubled. The province of British Columbia in Canada introduced a carbon tax in 2008, slightly raising the price of vehicle fuel and home energy. The UK government has legislated for all new homes to be zero-carbon by 2016.* Germany installed a record 3,800 MW of solar panels in 2009 (by comparison, 500 MW were installed across the entire US), and is aiming for 100-per-cent renewable electricity by 2050. China, the world's biggest energy user, is aiming for 20 per cent of its electricity to be renewable by 2020.

All of these examples were the result of significant public pressure, with governments often being forced reluctantly to put climate-friendly laws in place. This is despite the fact that – even with the recent rise in climate denial – the majority of the public still believe that climate change is real and serious, and want their governments to take action. A BBC poll of 24,000 people in 23 countries in December 2009 found that 61 per cent wanted 'state spending on climate change, even if it hurts the economy', while just 29 per cent opposed this.[6] Even in the US, a supposed hotbed of denialism, when an August 2010 poll asked

---

* They define a zero-carbon house as having 'zero net emissions of carbon dioxide from all energy use in the home'. Quite what that will mean in practice isn't yet clear...

whether 'the government should regulate greenhouse gases from sources like power plants and refineries in an effort to reduce global warming', 60 per cent of respondents supported this and just 34 per cent opposed it. When Australian Prime Minister Kevin Rudd failed to keep his promise to introduce climate legislation, he was forced to resign. But despite all of this, Northern governments are still dragging their feet over the climate issue.

Why aren't they doing more?
1) Oily palms
Transnational corporations have grown hugely in wealth and power over the last few decades. Of the world's 150 biggest economic entities, 95 (63 per cent) are businesses.* This gives them unprecedented influence over the world economy, and the ability to spend enormous amounts of money to secure laws and regulations that suit their interests. For example, oil and gas groups spent $354 million on lobbying governments between 1998 and 2004.** From 1999 to 2010, coal, gas and oil companies gave over $114 million to US politicians alone.[7] The companies probably consider this money well spent, considering the billions they received in subsidies and tax breaks during that time.

This money influences politics in two different ways. First, it helps to get polluter-friendly politicians elected – in nine out of ten US Congress elections since 1992, the candidate who spent the most money was the winner.[8] US Senators who voted against vehicle efficiency standards in June 2010 took, on average, 2.5 times as much fossil fuel money as those Senators

---

* Wal-Mart, Exxon, and Shell all have higher incomes than Indonesia, Norway or Denmark. General Motors is bigger than Ireland, and Toyota has more income than Portugal. See http://nin.tl/elXlYF
** President George W Bush was the single largest recipient of this cash, with $1.7 million flowing into his campaign funds.

who voted in favor of fuel efficiency. This pattern is repeated in parliaments around the world (although the difference is often less stark than in US politics).

Second, fossil fuel money distorts the political process by paying for professional lobbyists, advertising and PR campaigns. Environmental organizations campaigning for the 2010 US Climate Bill were reportedly outspent ten to one by their corporate opponents. As a result, the proposed legislation – never particularly strong in the first place – was repeatedly watered down and then finally dropped.

2) It's the stupid economy

Corporate influences aren't just about direct lobbying, though. We've set up our economies to value short-term profits and economic growth more than people's health and livelihoods, or humanity's long-term survival. This gives large businesses and financial institutions huge influence over governments, because they can deliver all sorts of shiny short-term economic benefits – jobs, income, tax revenue – to politicians, in return for lighter regulation. Never mind that the supposed benefits of all these new coal mines, factories or shopping centers come with far bigger hidden costs like local air pollution, increased waste, more traffic on the roads, a net loss of jobs as small local shops are shut down, and of course increased carbon emissions.

Australian coal-mining companies don't need to pay politicians directly to get permission to open new mines, because mining makes up 10 per cent of the country's GDP and 40 per cent of its exports. In 2008, tar sands extraction was responsible for 6 per cent of Canada's GDP and 26 per cent of its exports; it's therefore unsurprising that the Canadian government was one of the most obstructive at the 2009 Copenhagen conference, opposing binding targets and demanding stricter regulation of poorer nations. As a result, it won the Climate Action Network's 'Colossal Fossil' spoof award for the most anti-climate delegation at the talks.

3) What democracy?

As governments have been getting closer to big business, they've been getting further away from their voters. In the US, 80 per cent think that the country is run by a few big interests, and 94 per cent think that politicians don't pay attention to public opinion.[9] Between elections, our governments routinely ignore their manifesto pledges and allow their policies to be defined by their industry allies or 'scandals' whipped up by the mainstream media. Driven by short-term electoral cycles, most politicians are reluctant to think in the long term, or to take actions that might be unpopular with the media, business or the markets. This creates a narrow range of 'acceptable' political debate, which is increasingly out of step with the wishes of the public. Smaller parties (such as Green parties) that refuse to take corporate funding are out-spent in campaigns and struggle to win the attention of the mainstream media. All of this leaves political parties looking more and more similar to each other when it comes to climate policy. At this moment, none of the big parties in any wealthy Northern country are offering policies that would genuinely deal with climate change.

This isn't what democracy is meant to be about. We elect governments to look after our shared interests and to act for the common good. We need rapid action to deal with climate change, and politics in its current form – a few political parties fighting over a small patch of ground marked out in advance by their corporate sponsors, and ignoring the wishes of most of the public – isn't going to get us there in time. By way of contrast, countries with more participatory forms of democracy – such as Bolivia and Brazil – have seen genuinely reforming governments acting in a way that benefits the majority of their people rather than large companies. Of course there are still big problems in these countries, but they offer us some very useful comparisons. How much influence do *you* feel you have over your elected representatives when there isn't an election going on?

Governments of the Global South

What have they contributed so far?

Things are a bit different in the Global South. It's actually in the immediate national interest of many Majority World governments to tackle climate change, because their populations and economies are already being hit by its effects. As a result, coalitions of Southern governments became increasingly vocal throughout the UN Framework Convention on Climate Change talks, calling for strict, binding carbon reduction targets on the Northern nations that had created most of the problem, and requesting funding to help them lift their populations out of poverty in a low-carbon way.

However, despite these fine words, things on the ground are still generally heading in the wrong direction. Governments of the Global South rarely refuse permission to corporations wishing to extract fossil fuels or carry out unsustainable logging and intensive agriculture. The governments of oil-producing states such as Nigeria and Burma violently suppress their populations if they dare to stand up against drilling operations. Farmers are routinely removed from their lands to make way for polluting mega-projects like mines or dams, and indigenous forest peoples are displaced in favor of logging companies. Countries like China, India and Brazil who are currently experiencing rapid economic growth are powering it largely with fossil fuels, particularly coal.

There are exceptions. In the Indian state of Orissa,

hundreds of villages are now powered entirely by solar energy, with plans to increase this total to 2,000 villages by 2012. China has been closing its older coal plants, and aims to reduce the proportion of its electricity that comes from coal from 70 per cent to 50 per cent. South Korea is spending $45 billion to jump-start its economy, of which $36 billion – 80 per cent – is earmarked for energy efficiency, renewables and cleaning up pollution.

More excitingly, Ecuador looks set to be the first country voluntarily to leave a decent chunk of fossil fuels in the ground. In 2008, the right of nature to not be exploited and destroyed was enshrined in Ecuador's constitution (which was then approved in a national referendum) – a world first. In line with this commitment, the Ecuadorian government signed a binding agreement in 2010 not to dig up the estimated 850 million barrels of oil under the Yasuní National Park, thus also preserving at least 200,000 hectares of rainforest – so long as wealthier governments are prepared to pay them compensation. For $350 million per year over a ten-year period, (about half the estimated value of the oil), Ecuador will leave the oil underground in perpetuity, preventing over 400 million tons of CO_2 from being released to the atmosphere. A number of European governments have expressed interest in supporting the scheme – it would represent a tiny fraction of their annual national budgets. If this plan comes off, it could be a genuine example of Northern nations accepting their historic climate change responsibility, and a vital step towards leaving the remaining fossil fuels in the ground. However, the Northern governments seem to be developing cold feet, and the fate of the project is hanging in the balance.

The government of the Maldives, seeing that climate change threatens the very existence of their low-lying islands, have launched a bold plan to shift to a zero emissions economy by 2020, generating all their power from wind, sun and coconut husks. The Bolivian government, meanwhile, has taken a far-sighted political stand on the issue, calling for a rapid global phase-out of fossil fuels, the holding of temperature rises at 1°C and the support of the land rights of indigenous peoples as a way to protect the world's forests. Bolivia hosted an alternative global climate summit at Cochabamba in April 2010, bringing together grassroots movements from around the world to demand genuine solutions to climate change. Their political position is genuinely inspiring, but their on-the-ground practices are still far from perfect – their economy is still heavily reliant on oil and gas, and their CO_2 emissions rose by 16 per cent between 2006 and 2007.

Cuba, on the other hand, stands as a fascinating example of what a low-carbon economy might look like. After the collapse of the USSR in 1991, Cuba was forced to survive without cheap oil, thanks to a US embargo. The country shifted rapidly to organic agriculture, and wind and solar power are spreading across the island. Communities came together and worked out low carbon solutions to their most urgent shortages; the city of Havana now has so many productive gardens and allotments that it is self-sufficient in fruit and vegetables. As a result, many people in Cuba now have a better quality of life than in the days of cheap Russian oil.

Why aren't they doing more?

The most obvious barrier is money. For Southern nations to shift to a sustainable development path, Northern nations will need to accept their historic climate responsibility and inject some serious cash into the process, helping poorer countries switch to renewable energy as fast as possible.

The problem runs deeper than this, however. The

Why haven't we fixed it yet?

Southern nations aren't poor by accident – they've had their economies aggressively reshaped by Northern governments, companies and institutions, to supply cheap food, fuels and materials to businesses and consumers. If we're going to tackle climate change, then these practices need to stop.

The Southern governments themselves aren't completely blameless – many are corrupt, undemocratic or even dictatorial. However, the worst governments (Nigeria, Democratic Republic of Congo and Burma spring to mind) are often the legacy of countries hollowed out by European colonial rule and left with crippled economies under the control of a wealthy few. In a number of cases, Northern governments have even been complicit in removing popular democratic governments in favor of more export-friendly regimes (check out the histories of Iran, Nicaragua, the Democratic Republic of Congo and Honduras for examples). Many authoritarian regimes enjoy lucrative trade deals with Northern companies and cordial relationships with wealthy governments, all of which helps them to remain in place.

Southern governments are also under huge pressure to keep their unsustainable exports going, in order to pay off their historic (usually unfair) debts to Northern banks and governments, and due to pressure from international financial institutions like the International Monetary Fund (see below). Much of the money available for development projects, through international aid or World Bank loans, comes with strings attached that require the recipients to spend it on extractive projects, polluting infrastructure or unsustainable farming. The renewable technologies needed for zero-carbon development are often protected by patents and intellectual property rules, enforced by the World Trade Organization (see below), that makes them exorbitantly expensive for cash-strapped governments.

The private sector

What have they contributed so far?

Businesses have played a powerful role in the climate change story for a long time – and, sadly, it's mainly been a negative one. Back in the early 1900s, companies like General Motors and Firestone aggressively bought up public transport systems and shut them down, thus eliminating competition to the motor car. Over a hundred commuter rail systems and thousands of kilometers of railway lines across the US were stripped away and replaced – thanks to concerted corporate lobbying – with a massive, publicly funded highway system.[10] The US love affair with the motor car was really more of an arranged marriage.

Similarly, our current consumerist culture wasn't accidental or natural – it was invented by corporations and governments. After the Second World War, politicians and manufacturers found a common cause: the promotion of economic growth and consumption. It didn't matter whether this consumption was necessary, or added anything to people's lives – so long as people were buying more and more each year, then governments and manufacturers could both reap the short-term benefits of economic growth. The public needed to be convinced that they needed all this extra stuff, which is where marketing companies came in – spending on advertising in the US increased by more than a billion dollars a year. Companies realized that it was more profitable to produce cheap products that had to be frequently replaced rather than more durable items, and so things started being designed to break and run down.

Of course, having access to a wide range of new goods, services and foods did improve the lives of many Northern consumers – but at a heavy price, by locking Southern nations into unfair export arrangements, and setting the world on a high-consumption path towards climate disaster. According to US researchers, happiness in Northern countries has remained pretty much the same since the 1950s,[11] so all of the extra consumption after that point hasn't even made us any happier.

The influence of business has played a big part elsewhere in the climate change story – the shift from smaller-scale sustainable agriculture to more profitable intensive farming; the expansion of aviation in the late 20th and early 21st centuries; the movement of much Northern manufacturing to the Global South. In each case, the companies involved actively pursued the most profitable path, even though it was also a more polluting one.

Nowadays, many companies are keen to tout their green credentials; but under the surface it seems that little has changed. BP claimed to have gone 'Beyond Petroleum' in 1997; yet at the start of 2010, 94 per cent of their investments were still in oil and gas. Car companies like General Motors and Toyota sell a few flagship 'green' cars (like the Prius, which still burns oil but gets more kilometers to the liter), but make most of their money shifting millions of polluting gas-guzzlers. Supermarkets stock an aisle or two of eco-products, but fill the rest of the store with intensively farmed foods and cheap plastic toys. All of these companies are involved in lobbying against environmental regulations and for bigger markets for their products.

Some medium-to-large companies seem to be genuinely attempting to be ethical. Many insurance companies have realized that it's in their interest to avoid future climate disasters, have actively spoken out on the issue and are trying to reduce their own emissions. Renewable power company Solarcentury was founded specifically to fight climate change; its Chair Jeremy Leggett talks openly about the difficulties of trying to be sustainable in an economy based on cheap oil and endless growth. The British soap company Lush has banned palm oil from all its products (palm oil plantations are causing mass deforestation in Southeast Asia) and gives thousands of dollars to radical climate activists every year. However, companies like this are very much the exception, and even these pioneering firms aren't zero-carbon – both solar panels and cosmetics take energy to manufacture and transport, much of which still comes from fossil fuels.

Why aren't they doing more to help?
1) False profits

It all comes down to profit. Company directors have a legal duty to maximize the profit for their shareholders *above all else*. It is actually unlawful for public companies to consider the wider interests of society and the environment if that conflicts with making money.[12]

If their most profitable products or services actively help to solve climate change, then that's fine, they just need to sell more of them – but this is currently true for very few companies, and for no big corporations. For the rest, there are only three ways to be climate-friendly that don't conflict with profit-making:

- Climate initiatives based on efficiency that save money for companies by using less energy or raw materials.
- Climate initiatives that improve the company's brand, thus increasing sales.
- External factors (government regulation, public protest) that make it more expensive to go down the polluting route and thus force the company to cut its emissions.

Many companies have achieved great things on the efficiency front, by making lighter products, reusing waste, and cutting unnecessary energy use. However, every efficiency drive has a limit, a point at which no more savings can be squeezed out without increasing costs. Meanwhile, the efficiency savings that have been achieved haven't helped us much. Between 1970 and 2000, wealthy nations increased their CO_2-per-dollar efficiency by up to 40 per cent – but their emissions kept on rising, because these savings were all canceled out by increased production and growth.[13] There's also a problem called 'The Jevons Paradox' – reducing the amount of energy or materials that a company buys slightly reduces the demand in the wider economy, which reduces the price, which encourages someone elsewhere to buy more of it, thus wiping out the saving. Voluntary efficiency won't get us anywhere near the carbon cuts we need.

Brand improvement is even less helpful. Corporate Watch's excellent 2006 report 'What's Wrong with Corporate Social Responsibility' details how companies can get all the brand benefits of being 'green' by only taking very small actions. By making minimal changes to their product's ingredients, running slightly fewer trucks or giving a bit of cash to an environmental charity, companies can declare that they're 'now even greener' or 'eco-friendly'. This allows them to boost their green credentials, deflect campaigners' criticism and increase their sales without having to make any real changes to their core business practices. Here's the kicker: because these tokenistic efforts allow them to get the benefits of green branding cheaply, and because the law forces companies to put their profits first, it actually becomes unlawful for them to make the real, more expensive changes that are needed to prevent climate disaster. If a bit of green window dressing gives pretty much the same brand boost as a wholesale reinvention, then the company *must* plump for the window dressing.

The corporations that need to make the biggest changes – the ones digging up fossil fuels, building cars, manufacturing cement and chopping down forests – have a much deeper problem that can't be changed by efficiency drives or slightly greener brands. Their very core business – the way they make most of their profits – is fundamentally unsustainable. Many corporate executives may be perfectly happy with this, or in denial about it, but even those employees who genuinely want their company to change its ways are stuck in a conundrum: the law demands that corporations must always do whatever is most profitable for their shareholders, whatever the consequences might be for the wider world. It simply isn't possible for them to stop drilling for oil or making

---

Exxon vs BP – a story for our times

In 2009, the world's two biggest oil companies were Exxon and BP. Their actions over the last 20 years are excellent examples of the two main methods that polluting companies have used in response to climate change: denial and greenwash.

Exxon has been at the forefront of the denial industry, pouring millions of dollars into think-tanks, rogue scientists and politicians willing to stand up and declare that climate change isn't a problem. Despite finally, grudgingly, admitting that human-made climate change is happening, and promising in 2008 to stop funding deniers, the British *Times* reported in July 2010 that the company was still at it, and had given another $1.5 million to anti-climate-science groups.

BP took a different route. When oil companies came under fire for unethical and polluting practices in the 1990s – especially when Shell was implicated in the death of Nigerian activist Ken Saro-Wiwa in 1995 – BP changed its branding and claimed to have gone 'Beyond Petroleum'. It bought up a number of solar power companies and was hailed by many environmental groups as a green leader. Its CEO spoke at lots of conferences about the need to move beyond oil, and the company played an active role in pushing corporate-friendly climate 'solutions' like carbon trading. Some 13 years after the rebrand, the world was reminded how little had really changed when a technical failure on BP's Deep Horizon oil rig poured millions of liters of oil into the Mexican Gulf. Despite all its fine words, 94 per cent of BP's business was still in fossil fuels, and at the time of going to press it is pushing ahead with a plan to start exploiting the Canadian tar sands.

Imagine that someone has redirected a sewage pipe into your living room. Understandably, you're demanding that they remove it, and have put in a complaint to the local council. The Exxon response would be to deny that the sewage pipe exists, bribe the council to turn a blind eye, send in a team of scientists to explain that raw sewage is actually good for you, and keep the sewage flowing. The BP response would be to assure you that they're working on the problem, get funding from the local council to develop slightly cleaner sewage, send you some pot plants to brighten up your living room, and keep the sewage flowing. ■

inefficient cars, so long as it makes them more money than the alternatives.

It's that third category – external factors – that holds the greatest hope for changing company behavior. Well-enforced laws can make it more expensive for a company to pollute than to clean up its act. Public protest and direct action can damage a company's reputation or increase its costs, making polluting policies more trouble than they're worth. Sadly, we don't yet have enough proper regulation or protest to push corporations onto a sustainable path.

2) Growing pains

Meanwhile, there are thousands of smaller, alternative companies out there trying to be genuinely low-carbon, but they have much less power and influence than their big unsustainable rivals. It isn't the independent stores, family-owned farms or small green printing companies who get to rewrite the global trade rules or lobby governments for huge subsidies.

The only way small 'green' businesses can survive in an economy based on constant growth is to grow as rapidly as possible. They come under greater and greater pressure to adopt the cut-throat business practices of their rivals. This often means cutting costs, lowering standards, and increasing their carbon emissions. This process is not inevitable, but so long as the majority of companies are doing things cheaply rather than cleanly, genuinely low-carbon alternatives will struggle to compete.

International Financial Institutions

What have they contributed so far?

The International Monetary Fund (IMF), the World Bank and the World Trade Organization (WTO) were all set up with the stated aims of keeping the global economy 'stable', and encouraging 'free trade' between nations. Whatever noble ambitions may have been involved in their founding, in reality they have helped to keep most of the world in poverty, forced Southern nations into unsustainable development paths, and poured money into polluting industries and activities.

The IMF was founded after World War Two, to supply countries in economic downturn with extra funds to stop their currencies from collapsing. Its role has changed and developed over the years, and its main job now is as a financial enforcer, ensuring that countries repay their international debts. It forces indebted countries to restructure their economies, maximizing their short-term income and decreasing their public spending, as this will supposedly help them to pay off their debts more quickly. This usually has dreadful effects on the countries involved. For example, Costa Rica was 'structurally adjusted' by the IMF in the 1980s. Industrial banana and cattle production were ramped up, leading to deforestation, water pollution, land degradation, and the loss of farmland for subsistence agriculture. When the IMF forced Ghana to focus all its efforts on cocoa production, it led to a sharp drop in the global chocolate price and an economic disaster for the Ghanaians, many of whom then turned to tropical logging to make ends meet. A similar process happened in Vietnam with coffee production, and as a result Vietnam's forests are close to being wiped out. These IMF programs don't even succeed in increasing countries' incomes – the Fund's own research has shown that there is no significant difference in the economic performance of countries that went through 'structural adjustment' and those who didn't.[14]

The World Bank was originally meant to lend money to help war-ravaged countries rebuild their infrastructure. Nowadays, it is the biggest single source of finance for infrastructure projects in the Global South, and so has a huge influence over what gets built in these countries.

Why haven't we fixed it yet?

Sadly, despite many grand announcements about its various ethical policies, the World Bank rarely seems to take environmental or social factors into account when picking what projects to fund. In the 1990s, its projects were responsible for an estimated seven billion tonnes of CO_2 per year. In 2006, more than three-quarters of its lending for energy projects went into fossil fuels, and the total has been increasing year-on-year. It has been a major driver of coal mining and burning in China, Russia, India, Ukraine, and Pakistan, to the tune of billions of dollars. It pours huge amounts of money into oil wells, pipelines, and deforestation projects all over the world, and encourages other lenders to do the same.

The WTO is a painfully ironic organization. In the name of encouraging 'free trade', it enforces global rules that strictly control world trade, making it work in favor of wealthy nations and corporations. Member countries face tough sanctions if they try to restrict the sale or import of any goods, for any reason. 'Trade restrictions' are defined so broadly by the WTO that they can include environmental, health and human rights laws. The WTO also sets and enforces rules on patents and intellectual property. This relates to climate change in two important ways. First, it makes it more difficult for any country to pass a law banning the sale of high-carbon goods, such as steel made with coal, or oil from the tar sands. The countries producing those goods could use WTO rules to complain that they were being unfairly discriminated against, and the WTO would have the power to overturn the domestic law. This has happened in the case of other environmental laws, such as the 1990 US Clean Air Act which tried to restrict the import of the most polluting types of gasoline. Second, its strict patent laws make it more difficult for Southern countries to access affordable renewable technology. Regional and bilateral trade agreements such as the North American Free Trade Agreement (NAFTA) have had similar impacts.

Why aren't they doing more to help?

It's very telling that, in all of the above cases, the only real winners have been Northern corporations, who have benefited from new markets, increased profits and a supply of cheap raw materials. The real problem with the IMF, World Bank and WTO is that they are fundamentally undemocratic and unaccountable organizations, controlled by the people who put the most money into them – wealthy Northern governments and industries. Decisions are taken behind closed doors, and the public don't have any input at all. It's hardly surprising that their activities have ended up helping those wealthy interests at the expense of people and the climate.

Economists

What have they contributed so far?

Mainstream economics is dominated by 'neoclassical' theory. Put simply, this kind of economics believes that the best way to deliver wealth to people is through maximum growth, and the best way to achieve this is by allowing businesses to extract, refine, manufacture and trade as much as possible, as fast as possible. Anything that gets in the way of that – such as public control over resources, government-run public services like health and education, environmental regulations or labor standards – are seen as 'barriers to the free market' and should be removed or privatized.

This kind of economics has failed. To create a fairer world and avoid climate disaster, we need more regulation of polluting companies, not less. We need people and communities to have more power than the markets. Economists' obsession with growth and short-term profit above all else has helped to bring us to the brink of climate crisis, by stripping out the planet's natural resources, pumping pollution into the air, and exploiting the cheap or unpaid labor of the world's most impoverished people. Although it has improved the lives of many people, those benefits have been massively concentrated amongst a minority of the world's population. According to the 'Growth Isn't Possible', report from the New Economics Foundation, 'Between 1990 and 2001, for every $100 worth of growth in the world's income per person, just $0.60, down from $2.20 the previous decade, found its target and contributed to reducing poverty below the $1-a-day line.'[15]

By 'growth', economists mean an increase in Gross Domestic Product (GDP). This is a measure of economic activity – all it tells us is how much money has been spent on goods and services. It's therefore a very crude measure – it tells us nothing about whether those goods and services have brought any real benefit to anyone's lives. When a flood or a storm hits, the GDP goes up because of all the extra economic activity involved in paying for all the rescue teams, cleanup, rebuilding, hospital bills and funerals. However, there's an unspoken assumption that increased GDP is always good. To keep GDP growing, we need to keep making, buying and selling more and more stuff every year. This obviously requires more materials and energy each year – which directly contradicts our need to reduce our energy use in order to tackle climate change. Endless growth only ever seemed possible because we had access to so much cheap fossil energy, but that's no longer the case. We need to leave fossil fuels in the ground.

Why aren't they doing more to help?
There are plenty of more realistic economists out there proposing different ways of running the economy, and their voices are becoming louder. For example, in 2009, Elinor Ostrom won the Nobel Prize for Economics, for her work showing that local communities can successfully manage common resources without the

need for government control or privatization. Herman Daly, EF Schumacher and other key thinkers have been developing 'steady-state' or 'balanced' economic models, based on sustainable production rather than endless growth, for decades now. They have shown that it's perfectly possible to have secure jobs, pensions and savings without growth. In fact, a balanced economy should be more secure, because it's no longer based on unsustainable rises followed by inevitable crashes, and the concept of a 'recession' would no longer exist.[16]

Unfortunately, governments and business are still locked into the idea of endless economic growth, even though it clearly makes no sense – how can we grow for ever on a finite planet? The problem is, growth has worked for a long time to deliver short-term visible benefits; the downsides are less obvious and further away. People talk about economics as though it's something scientific or technical, but in reality it's highly political – there's nothing natural or inevitable about this kind of economic system. We were purposely taken down this path by politicians and business leaders. Other paths are available.

Climate campaigners

What have they contributed so far?
Climate campaigners' biggest success has probably been the huge increase in public awareness and understanding of climate change. There have also been all kinds of local victories, all over the world – fossil fuel developments and deforestation projects have been halted, roads and runway expansions have been blocked, renewable energy and efficiency projects have been launched. Community groups have created amazing local and low-carbon food, energy and transport initiatives, and actively educated themselves and those around them. Green political parties have had some success, particularly in Western Europe

where they typically receive between 5 and 10 per cent of the popular vote and have helped to raise the profile of climate issues within mainstream politics (although they have also been criticized for betraying their principles when in government in Ireland and Germany). It's safe to say that every piece of climate-friendly legislation enacted by governments wouldn't have happened without public pressure.

Some important links have been made. Big development charities now recognize the importance of the climate issue and have become much more vocal about it. Historically, polluting businesses and governments have typically tried to pit workers and environmentalists against each other (by arguing that environmental measures would mean fewer jobs), but there are signs that labor activists and climate campaigners are starting to find common cause. For example, several UK Trades Unions have teamed up with the Campaign Against Climate Change group to demand government investment in a low-carbon economy to create 'One Million Climate Jobs'.

But despite all of this, CO_2 emissions are still rising.

Why aren't they doing more to help?

Large environmental NGOs – from more 'conservative' groups like the Sierra Club and League of Conservation Voters to more 'radical' groups like Greenpeace and Friends of the Earth – need to persuade their supporters that they're living up to their donations and producing results. This creates a strong pressure to work together with businesses and governments to achieve small changes in policies, which can then be announced as significant victories. Green groups have ended up applauding companies for making tiny changes, and not challenging them on all the rest of their polluting activities. It's become particularly bad in the US, where groups like Conservation International and The Nature Conservancy even take funding directly from polluters.

Many campaigners have also been too hung up on persuading people to make small lifestyle changes. Of course, using less energy at home, cutting our car use and buying a bit less junk are all positive things to do, but they won't create the scale of change we need – we need to restructure our whole economies, not just sort a few extra jars for the recycling. This has also led to an impression that fighting climate change is all about using less, giving things up or taking on extra household tasks. This is the opposite of inspiring. In reality, avoiding climate disaster is about working together with loads of amazing people to shift to a low-carbon world, which in many ways will be a better and fairer one.

At the other end of the scale, more radical activists haven't always been brilliant at explaining their position to the public, often coming across as preachy or strident. Climate campaigners are also sometimes accused of being just as ideological as the right-wing deniers, and of using climate change as an excuse to pursue a left-wing, anarchist or anti-capitalist agenda. Campaigners need to be careful about this – it is true that people who believe that we need to challenge corporate power or restructure the economy for other reasons are more likely to be active on climate change, because it supports rather than conflicts with their existing beliefs. However, it *is* true that tackling climate change will mean major changes to our economy and challenging some of the basic tenets of capitalism like endless growth. This is what the science is telling us, but it's important to remember that this is a big leap for most people to make. We need to communicate it sensitively but also honestly.

Partly as a result of all this, climate campaigners have not succeeded in turning most people's awareness into action – even though we accept it's happening, many of us are still in denial about the seriousness of climate change, and its relevance to our lives.

Why haven't we fixed it yet?

Luckily, things are starting to change. Frustrated NGO workers, sick of decades of slow progress in the face of a climate emergency, are challenging their employers to be more assertive. Alliances are being formed between climate campaigners, development groups, health and poverty campaigners, labor activists and frontline communities. Grassroots groups and campaign networks like 350.org, the Climate Camps and Climate Justice Action are training and supporting activists to take nonviolent direct action against polluters. This raises the profile of the issue and the seriousness of it in people's minds, pushes the debate rapidly forward, and can lead to polluting projects being delayed or dropped. Campaigners are now more honest in explaining the scale of change required, and are getting better at using language that engages and inspires people rather than depressing them or putting their backs up. None of this can happen too soon.

Frontline communities

What have they contributed so far?
The Saramaka people of Suriname won a major legal victory to protect their forest from the logging industry. Campaigners in Gabon suffered eviction and imprisonment but succeeded in protecting thousands of hectares of rainforest from mining. Gas pipelines and coal plants have been blockaded in India. Around 50 indigenous groups in Peru have successfully blocked the destruction of their forest lands, despite violent suppression by the state police. First Nations communities in Canada are using protest tactics and legal challenges to prevent tar sands expansion. These are just a few recent examples. We rarely hear about these campaigns here in the North, but this is the real climate change frontline.

Southern activists were instrumental in the protests that prevented the WTO from extending its destructive powers, and joined thousands of Northern campaigners in Copenhagen in 2009 in an attempt to lay siege to the climate conference. Frontline communities are well aware of the link between their local struggles and the threat of climate change – many of them are already feeling its effects, or the effects of failed 'solutions' like biofuels. In April 2009, indigenous representatives from all over the world held their own climate summit in Alaska, and put together a declaration asserting their rights to defend their lands from fossil fuel development and deforestation. First Nations activists from Canada have now brought their fight to Europe and are linking up with climate campaigners across the continent. International links are being formed that offer real hope for the future.

Why aren't they doing more to help?
Frontline activists are often disenfranchised and shut out of official government processes. They frequently face far greater risks and penalties for standing up to state and corporate power than campaigners in the North. They have less access to financial resources and are often ignored by the international media. They are the people most directly affected by climate change, they are doing more than most to fight against it, and hold some important solutions to the problem. Despite this, frontline communities are often left out of the official climate change story here in the North.

The media

What have they contributed so far?
The mainstream media have played a vital role in informing the public about climate change, but they have done it in a frequently confusing, misleading and inconsistent way. As well as giving disproportionate coverage to deniers, they have tended to report on climate change in a sporadic and contradictory way.

Why haven't we fixed it yet?

While 'liberal' media outlets present us with sensational worst-case climate scenarios followed by cheerful lifestyle features on low-energy lightbulbs and eco-friendly shower heads, other newspapers and TV stations take an actively hostile stance on the issue. Reports of disastrous weather events only occasionally mention the probable link to global warming. It's usually reported as a political issue, with different 'lobbies' and 'factions' battling over environmental policy details, and little information about the underlying science. Brief bursts of good reporting (such as immediately after Hurricane Katrina in 2005) are often followed by periods of silence as the news agenda rolls on to another topic. A few crusading journalists provide pockets of useful information and good sense, but most seem to know or care little about the topic.

Luckily, there are also a plenty of alternative media outlets and websites where viewers, listeners and readers can encounter less filtered and more upfront reports about what's going on with the climate. Though these vary in quality, there are some great sources of information out there[17] – but sadly they attract fewer people than the commercial media.

Why aren't they doing more to help?

In *Flat Earth News*, Nick Davies explains how commercial pressures, staff shortages and the shortened deadlines of the 24-hour news cycle have turned journalism into 'churnalism' – investigative, informed and analytical reporting has been largely replaced with cutting-and-pasting from press releases and newswires.[18] The post of environmental correspondent is often seen as little more than a career stop on the way up the journalism ladder. Few journalists have any scientific training – in *Boiling Point*, Ross Gelbspan describes the amazement in a room of reporters when one of them described actually reading a scientific paper rather than just relying on the press release.[19]

Many editors are as much in denial about the climate issue as the rest of the population.

Advertising contracts and commercial pressures deter journalists from writing anything too critical of their financial backers – including car companies, travel agents and fossil fuel corporations. Other news outlets reflect the political preferences of their wealthy corporate owners and operators, and so are very unlikely to report honestly on a crisis that requires the reining in of corporate power.[20]

The endless roll of the news cycle is exactly the wrong sort of structure for reporting a slow, ongoing crisis like climate change. After a few reports, editors start to feel that the story has been 'done' and so ignore it for a while – or worse, print a piece by a denier in the name of 'balance'. In September 2010, former BBC correspondent Mark Brayne reported that the BBC had parked climate change in the category of 'Done That Already, Nothing New to Say'.[21] It's clear that we can't rely on the mainstream media to report the climate threat properly.

Common Threads

Difficult facts to work around
Climate change is hard for people to get their heads around, difficult to communicate clearly, and tends to have the greatest effect on the people who aren't causing it. Fossil fuels are incredibly powerful, useful and there's enough of them left to tip us into runaway climate change.

Things in our way that we might be able to change
- The distortion of politics by corporate donations, lobbying and PR campaigns.
- The huge subsidies paid to the fossil-fuel industry, that make coal, oil and gas appear artificially cheap.
- The way our economies are based on the idea of

endless GDP growth and the maximizing of profit above all else, while alternative economic models are sidelined and ignored.

- Our increasingly unaccountable and flawed democratic systems (and the total lack of formal democracy in a number of Southern countries).
- The unfair systems of international trade and debt that keep Southern nations poor.
- The lack of effective legislation to rein in polluting business activities.
- The piece of corporate law that forces public companies to place profit above all else.
- The undemocratic, unaccountable and destructive power of the IMF, World Bank and WTO.
- The lack of effective public pressure on governments and businesses to clean up their acts, and of a coherent movement building a zero-carbon society from the ground up.
- A popular culture based on consumption and a disconnection from politics.
- A mainstream media that's not taking the issue seriously.
- Not enough links between climate campaigners and activists working on other issues.
- A lack of support for frontline climate struggles.

These barriers may seem daunting, but they also give us an exciting opportunity. Tackling these issues will also help with many other important causes and campaigns, not just climate change, giving us the chance to find common ground with other groups and movements all over the world.

1 IEA/OPEC/OECD/World Bank joint report (2010) 'Analysis of the scope of energy subsidies and suggestions for the G-20 Initiative'. Available at www.worldenergyoutlook.org **2** A Morales, 'Fossil Fuel Subsidies Are 12 Times Support for Renewables, Study Shows', 29 July 2010, www.bloomberg.com **3** Environmental Law Instltute (2009) http://nin.tl/atv3RA **4** Institute for Sustainable Futures (2007) 'Energy and transport subsidies in Australia'. Available at www.isf.uts.edu.au **5** A Grice, 'Revealed: airlines' £10bn government fuel subsidy', *The Independent*, 9 Jun 2008. **6** BBC news, 'BBC climate change poll shows rising concerns'. 7 Dec 2009. **7** Oil Change International (2010), 'Dirty Energy Money Abounds in 111th Congress'. Report available at http://priceofoil.org **8** Americans for Campaign Reform (2008), 'Does money buy elections?' Policy paper available at www.scribd.com **9** Noam Chomsky, *Hopes and Prospects*, Hamish Hamilton, 2010. **10** P Hossay, *Unsustainable*, Zed, 2006. **11** National Opinion Research Center at the University of Chicago; Pew Research Center. **12** Joel Bakan, *The Corporation*, Free Press, 2004. **13** New Economics Foundation, *Growth Isn't Possible*, 2010. Available at www.neweconomics.org **14** Prasad et al (2003) 'Effects of Financial Globalization on Developing Countries' IMF Occasional Paper, as cited in Patrick Hossay, *Unsustainable*, Zed, 2006. **15** See footnote 13. **16** A good starting point to learn more about all this is http://nin.tl/bbfPIX **17** See, for example, www.indymedia.org www.zcommunications.org www.therealnews.com and www.newint.org **18** Nick Davies, *Flat Earth News*, Chatto, 2008. **19** Ross Gelbspan, *Boiling Point*, Basic, 2004. **20** www.medialens.org **21** J Romm, www.climateprogress.org 22 Sep 2010.

9 What is it going to take?

The truth today is stranger than fiction... The greatest social change in history is already starting to happen... Emergency action and Green New Deals... The problems with climate adaptation and carbon offsets... Why the only option not available to us is to carry on as normal.

HOW EXACTLY DID global climate politics turn into the plot of a dodgy science fiction movie? Scientists are warning of a grave danger to the Earth, but aren't being taken seriously. Politicians and business leaders carry on as normal and ignore the approaching threat, offering just a few token efforts to reassure the public. Some people are sounding the alarm, but their message isn't getting through. In this case, we're talking about climatic calamity rather than an alien invasion or approaching asteroid, but otherwise the similarities are uncanny.

The main difference is that Bruce Willis and a wise-cracking sidekick aren't going to turn up in a rickety space shuttle and rescue us all (or at least it doesn't seem very likely). The only hope now is that a collection of unlikely heroes can save the day against all the odds – and, yes, I'm afraid that's going to have to be us.

We're going to need a serious amount of people power to bring down the barriers from Chapter 8 and turn things around. It's a big task, and time is tight, but I know we can manage it. Here's why.

Not toast yet

History teaches us that big social change is often sudden and unexpected. There were similarly tough barriers in the way of banning the slave trade, ending apartheid in South Africa, and achieving independence and democracy for countries all over the world. No-one thought that women would get the vote, workers would win the right to weekends and paid holidays, black people in the US would gain legal equality or that the Berlin Wall would fall, until these things happened. They happened because people stood up against seemingly impossible odds and didn't give in.

We know that a zero-carbon world is possible, and that it can be done in a way that also lifts people out of poverty. Whether we're talking about the wealthier minority escaping the consumerist treadmill and rediscovering the things that really make them happy, or everyone else gaining control over their lands and livelihoods, climate change is an issue which can bring together many different struggles and unite us with visions for a better world.

This gives us the foundations for the most amazing and powerful movement the world has ever seen. Getting fully active on the real root causes of climate change will bring so many more people on board. This is about so much more than totting up tonnes of carbon – it's about reclaiming democracy, defending jobs and livelihoods, making our streets safer and our homes warmer, fixing the economy, fighting poverty, hunger and disease, standing shoulder-to-shoulder with oppressed people and pushing for global justice. With these kinds of goals we can build a global movement of citizens, farmers, workers, landless peasants, environmentalists, trades unionists, poverty campaigners, equality activists, indigenous communities, refugee groups and many others, under the banner of 'climate justice'. In fact, this is already beginning to happen.

People across the world have stood up against destructive developments, and won. Increasing numbers of people are ready to do what it takes to create this change; many are willing to put their bodies on the line and risk their liberty to achieve climate justice. Within this growing movement exists an amazing range of skills to create sustainable solutions from the bottom up, and people are learning how to share and spread those skills. We don't all need to agree on everything (which is lucky), because there is more than enough common ground for us all to work together, and become strong enough to create the shift to a zero-carbon world.

It's easy to forget just how many people would welcome the changes we need. Most of the world's population would probably agree that the North should reduce its consumption, switch to zero-carbon energy, help the South onto a clean development path and defend local people's sovereignty over land, food and forests. It is only the uneven distribution of power across the world that is preventing these ideas from becoming the accepted wisdom. With enough people, determination and compassion, we can start to take that power back.

This is going to be the most amazing, inspiring and unifying social movement that the world has ever seen. It's going to be difficult, and frustrating at times, but it's also going to massively enrich the lives of everyone who's a part of it. This includes you. You may not think of yourself as a campaigner: you probably have far too much else going on in your life suddenly to become some kind of full-time climate activist. That's fine – there are still loads of things you can do to contribute to this, and to be part of the most exciting and important social uprising of our lifetimes. I'll tell you how in Chapter 11.

The art of the possible

The global movement for climate justice isn't like a political party or an NGO. There will never be one shared manifesto, or leader, or strategy. Different groups, communities and networks around the world will work in different ways to push for climate justice. It will probably happen through a mixture of formal coalitions, loose networks, unspoken agreements, statements of principle, shared funds, acts of solidarity, and lots and lots of talking to each other. We'll need to recognize our political differences and create space to discuss and debate them, while standing firmly together wherever we have common ground.

I saw this approach work in practice at the Copenhagen protests in December 2009, when groups from the Climate Justice Action and Climate Justice Now networks came together to create some amazingly unlikely alliances. African farmers left their fields and traveled north to plan a conference center invasion in Denmark. Vegan anarchists from Europe were championing the rights of Indian fisherfolk and South American hunter-gatherers. Indigenous peoples from across the world had come to the lands of their former colonial oppressors, and were debating campaign tactics in the languages of European Empire. The mutual respect, political concessions and genuine idea-sharing on display put the highly paid negotiators in the official UNFCCC meeting to shame. We marched together on the conference center in our thousands, and succeeded in holding an Assembly for real solutions in the jaws of what seemed to be Denmark's entire police force. If we can keep on building these alliances, then the Copenhagen protests might just be a small taste of great things to come.

If this movement is to grow and succeed, we'll need lots of different tactics and strategies to get us to a zero-carbon future. Below are some suggested policies, tools and tactics that might be useful along the way. Different things on this list will doubtless appeal to different people. In reality, we'll probably need a mix of all of them plus some others that we haven't thought of yet.

Emergency programs and Green New Deals
We need a huge expansion of renewable power, sustainable public transport, mass home insulation and all the other stuff from Chapter 6.

What is it going to take?

Where communities can take control over their own energy and transport systems, there are all kinds of benefits to local wealth, democracy and livelihoods. The more decentralized this process can be, the better. However, time is not on our side. Renewable technologies are still expensive to install, and most will need to be built on a large scale in prime locations if we are to get enough renewable electricity flowing in the short time we have. New sustainable technologies need to be nurtured and encouraged with subsidies and supportive legislation.

This means that, like it or not, we need massive government action and investment. We should do as much as we can to put grassroots solutions in place ourselves, but our governments are the only place we're going to find the resources to make big enough changes happen in the time available, and in a way that's effective and fair. This is why many Northern campaigners are calling for their governments to implement a 'Green New Deal' to tackle climate change, peak oil and the economic crisis at the same time, through a huge investment in low-carbon infrastructure that would also create millions of jobs.[1] Other campaigners speak in terms of shifting our governments to a state of 'Climate Emergency', comparing this to the way in which Europe, Australia and the US completely and successfully transformed their economies for military production during World War Two.[2]

Northern governments complain about not having enough money for all this. But this is a global emergency. If you wake up to find that your house is flooding, you don't ignore it and set off to work because you're worried about losing a day's pay – you start shifting sandbags. Wealthy Northern nations can find the money if they really need it; we found the cash to rescue the banks, surely we can find enough to save the rest of us too? Even in cold financial terms, the costs of the mass global destruction threatened by climate change are much, much greater than the costs of avoiding it.[3] For example, switching to a zero-carbon Australia would cost something in the region of US$300-400 billion over 10 years. This is about 3-4 per cent of their current economic production, a far smaller percentage than the 30 per cent that was typically diverted by countries involved in the Second World War. Providing all of India – a sixth of the planet – with renewable electricity would cost about $20 billion per year over 15 years, equivalent to around three per cent of the US military and intelligence budget.[4]

Money could also be raised by stopping subsidies to the fossil fuel industries, more proportional taxes on banks, big companies and the wealthiest individuals, a clampdown on tax-dodging millionaires and corporations, or a tax on global financial transactions. Proponents point out that this level of public investment would create a major boost to Northern economies and save significant amounts of money by moving millions of unemployed people into work (thus reducing benefit payments and boosting tax revenues).

Northern governments have not yet risen to this challenge. It's not going to happen on the scale required – including a just transition for energy workers and financial support for the Global South – until we've done some serious work to clean up our political systems and reclaim our democracies.

Shutting off the carbon tap

All the renewable technology in the world won't help us if we're still digging up fossil fuels and ripping down the rainforests. We need to tackle both ends of this problem, and find ways to keep the oil in the soil, the coal in the hole, the gas in the crevasse and the trees… er… swaying gently in the breeze.

Around the world, campaigners and frontline communities are acting to block forest loss and fossil fuel extraction, transport and burning, with a combination of legal challenges and direct action. This adds hugely to the cost of these destructive projects and has scuppered them completely in many cases. Meanwhile, governments need to withdraw their huge subsidies and support from the fossil fuel industry, and start toughening up legislation to phase out these fuels entirely. They'll need to challenge high-energy industries head-on, probably with a combination of taxes, non-tradable quotas, and the outright banning of the most polluting activities (such as tar sands extraction, open cast coal mining, mountaintop removal and deep offshore drilling). Indigenous peoples' land rights and small-scale farmers' food sovereignty will need to be respected over and above the 'right' of corporations to make profits.

All of this could be made much easier by reforms to corporate law, removing the imperative for public companies to put shareholder profits above all else, and preventing corporations from being given the same legal status as people (as is currently the case).[5]

The IMF, World Bank, WTO and trade agreements like NAFTA will need to be radically reformed or scrapped entirely, giving Southern nations the freedom to shift away from destructive and unjust export industries. Global grassroots movements have already had much success in slowing the expansion of the WTO's powers. The US seems to be losing interest in the IMF, leaving the institution in an unusually vulnerable position. Northern governments hold the purse-strings of these organizations and could make big changes to them if they chose. The writer and campaigner George Monbiot points out that Southern governments could force these kinds of reforms by threatening to withhold their debt repayments to the North.[6]

Once again, we need a global movement with enough clout to create these kinds of changes and to ensure that they are enacted in a way that is fair – for example, a rise in the price of fossil fuels could hit the poorest hardest unless policies are in place to prevent this.

I will if you will
We need action all over the world, especially in the highest-emitting nations and in those with the most fossil fuels and forests. Multinational companies need multinational controls, so we'll need a good number of governments to clamp down on them for their regulations to be effective. Up till now, the best hope for guaranteeing global action has been the UNFCCC process; however, until Northern governments are ready to make some concessions to the South based on their historic responsibility, and start talking about effective and fair ways of cutting fossil fuel use, then no meaningful agreement is ever likely to be reached.

This is very unlikely to happen until Northern countries are pushed onto a sustainable path themselves by their own populations, giving them a major incentive to bring other countries with them. We can't wait for a global deal to force our governments to act; we need our governments to act first, or there'll never be a global deal.

Moving beyond annual emissions targets to an approach based on keeping fossil fuels in the ground could also help reinvigorate the process. Rather than endless haggling over annual cuts that rapidly fall behind the science, the UNFCCC could focus on practical schemes to stop coal, oil and gas from being dug up. A global system of taxes or permits at the point of extraction could be used to phase out fossil fuels and raise money for a global fund to provide sustainable technology to the South.[7] Tariffs could be used to reduce the import of fossil fuels to Northern nations, with the money distributed to the public or spent on energy efficiency schemes to offset any short-term rise in energy costs.[8] Countries could agree to firmly regulate polluting industries and ban the worst practices outright (from Arctic drilling to high-carbon cement production to the release of 'fluoro' greenhouse gases). However, these agreements would need to have a strong justice element and meet the needs of those most affected by climate change, fossil fuel extraction, deforestation and industrial agriculture. The only way to ensure this is for these frontline communities to have a strong voice in the decisions. All of this is a long way away from the UNFCCC process as it currently works.

A People's Protocol

Can we really create the scale of political and social change necessary to stave off climate disaster without a formal intergovernmental treaty? There

---

Adapting to climate change

Some people say we should spend less time cutting carbon and more time adapting to the effects of climate change.

We're already committed to a certain level of climate change, and so some adaptation will be absolutely vital. If we're serious about climate justice then the nations most responsible for causing climate change should be providing funds and technology to the people on the receiving end, to help them cope with rising sea levels and more serious floods, storms and droughts. Some adaptation is already under way in wealthier nations – for example, London is planning new flood defenses and many of Britain's roads are being resurfaced with heat-resistant materials in anticipation of hotter summers.

However, adaptation cannot be a replacement for reducing greenhouse gas emissions. 'Adapting' to more serious levels of climate change would involve coping with mass food shortages, the loss of dozens of major cities, finding new homes for hundreds of millions of people and countless deaths from starvation, conflict and disease. Runaway climate change could leave us with a largely uninhabitable planet.

Even if it were possible, adaptation on this scale wouldn't be cheaper or easier than cutting CO_2 emissions – even conservative estimates like the UK government's Stern Review place the costs of climate impacts far higher than the costs of prevention.

It's a bit like... saying 'it's just too much effort to hit the brakes, I'm sure my car can adapt to that brick wall'. ∎

is a possible alternative. The political declarations that emerged from the alternative conferences at Copenhagen in 2009 and Cochabamba in 2010[9] have a lot in common with a number of People's Protocols already drafted by Southern movements.[10] In a nutshell, they all call for a reduction in Northern overconsumption, the abandonment of fossil fuels for cleaner alternatives, a transfer of wealth and technology from North to South for climate mitigation and adaptation, the rejection of false market-based solutions and geoengineering, strong recognition of indigenous land rights, and local sovereignty over food, land, energy and water. They also – implicitly or explicitly – call for a different kind of global economy and politics, based on the needs of people and the environment rather than corporate profit and endless growth. International grassroots movements are already far closer to a meaningful, scientifically robust agreement than the UNFCCC is likely to be any time soon.

These agreements need not just be a list of demands to governments. They could be developed into a set of real solutions that we intend to put in place ourselves, using every tool available to us. A 'People's Treaty' of this kind could be a rallying point, a shared global agreement under which people around the world could plan their own actions and set their own goals. This could help to galvanize an international movement, stopping fossil fuel extraction and creating sustainable alternatives, while building the strength of our movement to the point where we can force governments to accept their responsibilities too.

Growing up

Continuing with GDP growth while also tackling climate change appears to be physically impossible. While it might be politically challenging to shift to a balanced, non-growth economy, I suspect we'll have

Carbon offsets

Carbon offsetting is where, rather than actually reduce your own carbon emissions, you pay someone else to (supposedly) reduce them for you. It exists in two major forms: voluntary offsetting and governmental offsetting as part of the Kyoto agreement.

Voluntary offsetting takes place when an individual or organization decides to pay some money to an offsetting company. That company will then plant some trees, give out low-energy lightbulbs or set up some other sort of eco-scheme (normally in the Global South), thus saving some carbon on your behalf. If you've just done something polluting, such as taking a flight, offsetting companies claim to be able to reduce just the right amount of carbon dioxide elsewhere to 'neutralize' your emissions.

Government offsetting takes place under the name of the Clean Development Mechanism (CDM), which is part of the Kyoto Agreement. It works in the same way as voluntary offsetting, but on a bigger scale – countries can 'buy' part of their mandated emissions cut by paying for a carbon-reducing scheme somewhere in the Global South.

There are various problems with this. Working out how much carbon will be 'saved' by say, installing some low-energy technology in a village in Bangladesh, or planting some trees in Guatemala, is pretty much a matter of guesswork. When the rock band Coldplay paid an offsetting company $50,000 to plant 10,000 mango trees in India to offset the emissions from their tour, the scheme fell apart halfway through; only 8,000 trees were ever distributed, and most of them didn't survive. Even if it works, offsetting in this way only reduces emissions some time in the

more luck changing politics than we will trying to change physics.

The idea that governments should be focusing on other things than GDP isn't a particularly radical one – a UK survey in 2006 found that 81 per cent of people supported the idea that the government's primary objective should be greater happiness and not greater wealth.[11] Making it work, though, will take some serious changes to our economies. Researchers like Herman Daly have done decades of work in this area. They have already come up with all kinds of suggested techniques for setting a level of annual production that is ecologically sustainable, and then ensuring that the wealth generated gets to where it's needed, via new rules and regulations for banks, future – but we need enormous cuts in CO_2 right now. We need to keep fossil fuels in the ground *and* support low-carbon development in the Global South. Offsetting suggests that a bit of the latter lets us carry on with the former.

Many offset projects have negative impacts on Southern communities. People in Uganda were removed from their land to make way for a 'carbon forest'. Inappropriate pine tree planting caused erosion and soil loss in Ecuador. The CDM has supported some highly dubious schemes, including agrofuel plantations in Brazil, methane-capturing schemes that have forced toxic rubbish dumps to stay open next to people's homes in South Africa, and slightly more efficient Indian factories that were going to be built anyway.

As people get the idea that they can buy off their guilt by giving a few dollars to an offsetting company or charity, they are more likely to keep on driving, flying and ramping up the thermostat with a lovely warm glow in their hearts. They are also less likely to take all-important political action if they think that some nice companies are somehow going to 'sort it all out'. Offsetting also allows governments to avoid even the small reduction commitments in Kyoto by chucking a bit of money at a (probably awful) project somewhere else.

It's a bit like... Punching someone repeatedly in the face, but then making it all OK by giving some money to a hospital in Kenya. At least, you think it's a hospital. It might be a cigarette factory. No, you're pretty sure it's a hospital. ■

businesses and markets. Because production and consumption in the North will be lower, and just focused on the things that bring genuine benefits to our lives, we should be able to work less, spend less and live more. Governments will no longer be able to pretend that growth will lift people out of poverty, and so there'll need to be a greater focus on fairness and reducing inequality. Research has shown that more equal societies tend to be happier and safer than unequal ones.[12]

No-one yet knows all the details of how this would work, or what it would take to get there – but our current economic system is keeping hundreds of millions in poverty and driving us into climate destruction, so it's got to be worth trying out some new ideas. Some of this change could happen gradually, some would require major and rapid shifts in the way our economies work. Lots of people will enjoy arguing about whether these changes are possible within a capitalist system, and whether we're talking about changing to something else entirely – a 'post-capitalist' economy. It would certainly no longer be capitalism as we know it, which has the drive for growth and expansion at its very core. Ultimately, the precise terms we use to describe it are less important than figuring out something that works.

It's going to be quite a journey, but a useful first step might be to get national and local governments, communities, businesses and other organizations to accept that GDP is not a good measure of real social progress, and to start experimenting with alternatives. We also need to give more prominence and credence to alternative economic theories and practitioners around the world, from thinktanks like the UK's New Economics Foundation to Latin American social movements trying out alternative economies and politics on the ground in their communities.

Is any of this realistic?

The policies outlined in this chapter may seem very far away from the current position of governments and business. You may be feeling that these things are never likely to happen, that they're just not realistic. In fact, they are far more realistic than the idea that things can carry on the way they are going.

Although things are now very urgent, we're still talking about at least a 20-year transition. Some changes in public attitude (like breaking North American car dependency and getting wealthy Europeans to holiday without flying) are going to take a while to achieve, but many other measures could happen relatively fast and be popular with the public (such as refurbishing homes and improving public transport). The removal of coal from our energy supply should make little difference to the vast majority of Northern consumers so long as alternative energy sources are put in place and energy efficiency measures are brought in to reduce the costs to the public. In China, Russia and India, meanwhile, millions of people are crying out for a shift away from this filthy fuel that poisons their air and blights their lands. The tar sands are already starting to provoke international outrage. Deepwater oil drilling has never been less popular, thanks to the Gulf of Mexico spill. The idea that our economies are there to provide ever-growing profits for banks and corporations has also fallen out of public favor since the economic crash. The real barriers to many of the changes we need are coming from businesses and governments, not from the population, and there are far more of us than there are of them...

Change is coming, whether we like it or not. Our current path, based on ever-increasing consumption of resources and energy, is bringing us up sharply against loads of different natural limits – not just climate change, but also collapsing fish stocks, degraded soils, declining mineral resources, acidic oceans and depleted freshwater supplies. Our choice is between sensible (but rapid) changes, under our control, to a zero-carbon low-impact society; or disastrous changes forced upon us by an out-of-control climate, conflict for dwindling resources and mass ecosystem collapse. There is no 'carry on as normal' option.

We should aim to get to a zero-carbon society as rapidly as possible because it's the only sensible and moral thing to do; but even if we don't get there as fast as we'd like, our efforts will still have made a difference. Every fraction of a degree of warming that we prevent will save many people's lives and homes, and protect species from extinction. Just so long as we hit the key 'red line' targets in Chapter 5, we'll still have a chance of avoiding irreversible climate disaster.

What is it going to take?

Our economies and political systems were made by people, and can be changed by people – so long as we can get organized and active, and can bring a lot more people on board, fast.

1 www.greennewdealgroup.org 2 www.climateemergencynetwork.org 3 M McCarthy, 'Lord Stern on global warming: It's even worse than I thought', *The Independent*, 13 Mar 2009. 4 D Spratt & P Sutton, *Climate Code Red*, Scribe Books, 2008. 5 Joel Bakan, *The Corporation*. Free Press, 2004. 6 George Monbiot, *The Age of Consent*, Flamingo, 2003. 7 As detailed by Oliver Tickell in *Kyoto2*, Zed, 2008. 8 As detailed in James Hansen, *Storms of my Grandchildren*, Bloomsbury, 2009. 9 See http://declaration.klimaforum.org and http://nin.tl/aHW2qX 10 For example, by the Peoples' Movement on Climate Change at http://peoplesclimatemovement.net 11 www.gfknop.com 12 Richard Wilkinson & Kate Pickett, *The Spirit Level*, Allen Lane, 2009.

10 What might a zero-carbon future look like?

The following is one vision of what life in a zero-carbon world with a balanced economy might be like. It is not a prediction, just a description of one possible future in an industrialized Northern nation.

IT'S A MORNING in early March. You wake up slowly, reluctant to leave your warm bed even though your well-insulated home is a perfectly comfortable 19°C. The ground source heat pump had been playing up, but now seems to be working fine. You should probably still get someone to take a look at it, but you're a bit too busy to sort it out today. You get up, get dressed, and breakfast on fresh bread from the local bakery. You step out into the street – the air is clean, and there's hardly any traffic. A group of kids cycle past on their way to school as you stand at the bus stop, waiting for the bus out to the farm.

It works like this: you spend a couple of days a week on the community farm, and in return receive enough food to cover most of your family's needs. The work is shared around, so everyone takes a turn helping with planting, harvesting, purchasing, accounting and everything else. It's a low-input farm, what used to be called organic in the old days – it takes quite a lot of labor, and is a hassle when it's cold or wet, but produces a decent amount of food for everyone involved, plus a bit extra to sell. Quite a lot of people choose to work on community farms – it's good exercise, and massively cuts down the food bill.

You spend two more days per week in your job. It's a fairly normal job, pretty similar to a lot of the jobs that were around back in the 2010s. Many things are different in this zero-carbon society with a balanced (rather than expanding) economy, but people still need doctors and teachers, plumbers and accountants, insurance underwriters and street cleaners. Overall, there's less manufacturing in the world but far more of the things we use are made in this country, so plenty of the people in your town are involved in making or selling things – sometimes both. There are far more small independent workshops around since so many of the giant stores with their inefficient, transport-hungry, centralized supply chains went bust.

Work is generally a lot more pleasant these days – the wages are fairer, the buildings are comfortable, the commutes are short, and the whole atmosphere is a bit less frantic and cut-throat. Without every business desperately struggling to expand at all costs and trample its rivals, there's less pressure to overwork the staff or skimp on health and safety. Businesses still compete for custom, but it's about providing high-quality goods and services rather than slashing costs.

Best of all, because the economy's no longer based on producing and consuming as much stuff as possible, everyone can work a bit less. Thanks to your super-efficient home and appliances, your energy bills are tiny (most of your town's electricity comes from a wind farm 30 kilometers off the nearby coast, backed up by wave and tidal generators and a storage facility). Your transport costs are minimal: you walk and cycle most of the time, except for the bus out to the farm (which is reasonable, but you reckon it could be cheaper if the local council got their act together – you make a mental note to go to the next community meeting).

You also occasionally pay to use an electric car from the community car pool, for trips out to the countryside or buying new furniture, but that's not a big expense – far, far cheaper than owning a car! Most of your family's food comes from the community farm, topped up with whatever looks good at the local market and the occasional imported treat. Clothes aren't a big expense either – without the constant pressure of advertising and the fashion industry, most people are quite happy to keep clothes for ages, and buy second-hand. Things are generally made to last these days, and recycled clothing is *huge* – there are loads of boutiques that take in worn-out garments and transform them into snazzy new lines.

All in all, life is much less expensive than back during the consumer craze – you do two days of paid work per week and your partner does three, which is more than enough and means that childcare isn't a big expense either. Some people still choose to work full-time and have more spending money, but most prefer to have the extra time to spend with friends and family or pursuing their hobbies. Shopping is no longer a national pastime – most people prefer music, films, playing sport, watching TV (there's no getting rid of TV), socializing, or walking in the local woods. There's a lot more woodland around these days. A lot more walking, too. People are generally healthier than they used to be. They're also more connected with each other – the political movement to stop climate change involved taking quite a bit of power back from central government and into local communities. That's evolved into lots of local decision-making meetings, which can be a bit tedious at times but mean that everyone knows each other a lot better.

You're a little bit sad about the row of wind turbines on top of the ridge just outside town – you do think they could have been better sited – but they're definitely an improvement on the coal plant that used to sit in the valley. You also sometimes think about the flights you took as a child; how extraordinary it was to be able to travel so far, so fast. Still, it's much easier to take long stretches off work these days, so there's time to travel across continents by train, stopping off to explore different countries along the way. When you've saved up enough, you're even thinking about a family trip on an airship across the Atlantic... but that's still a bit of a long-term dream.

Life isn't perfect. There's still conflict, and crime, and injustice. Some people are still rich and others are still poor – although things are shared a bit more fairly now, and farmers, workers and indigenous peoples around the world have taken a lot of land and power back from the governments and corporations that used to oppress them. Politicians still tell lies and try to start wars; there was enough climate change to make disasters more frequent, and they affect some people more than others; there are still campaigns to be run and things to be sorted out.

But it's a better world than it was. The worst climate disasters were avoided, the world was made safer and fairer, and you were part of the generation which made that happen. Life isn't perfect – but it's pretty darned good.

* * *

You might think this sounds like a complete fantasy, or a possible reality. Either way, it's a lot more plausible than the idea that everything can just carry on the way it is. One way or another, for better or for worse, things are going to change. It's up to us to decide what sort of change we want to see.

11 Ten top tips to save the climate

*'Happiness is when what you think, what
you say and what you do are in harmony'*
Mahatma Gandhi

*'Just as the good life is something beyond the
pleasant life, the meaningful life is beyond the
good life'*
Martin Seligman, psychologist and happiness
expert

**How climate action can make you happy... Four things
for everyone to do... Ten key areas to work on... How
we can take control.**

THIS HASN'T BEEN the cheeriest book in the world.
I'm aware of that. I've tried to be honest about the
scale of the climate challenge without being *too*
gloomy, but it's still a lot to take in. It's not hard to
understand why many people, faced with these facts,
decide they'd prefer not to believe them, or to believe
that it's nothing to do with them, and someone else is
going to sort it all out.

This is a big mistake. Climate activists – people
who've grasped the reality of the problem and
decided to do something about it – are among
the happiest people I know. Study after study has
revealed that a major component of happiness
is having meaning in our lives, devoting time to
something larger than ourselves. Once you know
about climate change, once you realize how serious
it is, then absolutely the best way to deal with it is to
do something about it. In the words of Paul Connor,
co-founder of Climate Justice Fast!, 'We cannot be
at peace if our actions do not reflect what we truly
believe. But when they do, our spirits soar. Then
we're alive and free.'

Ten top tips to save the climate

I can testify to that. Some of the happiest times of my life have been spent on climate protests and action meetings, or standing up in front of audiences telling people about this stuff (often in rhyme). Yes, I know that's a bit sad. But it's true. Since I became active on climate change I've done all sorts of things I never imagined I'd do. I've met the most amazing and inspiring people from all over the world, people whose determination and compassion give me so much hope. I'm still worried about the future, but I believe that we can turn things around. The greatest cure for despair is action.

But what action? Well, this final chapter is here to give you some starting points. Below are four things that everyone can do. Then there's a list of 'Ten Top Tips' for beating climate change. Don't worry, you don't have to do them all – in fact, it'd be better if you just chose one or two to get started, and see how it goes. You can always come back and choose some other ones later.

Four things for everyone to do

Believe it. It takes everyone a while to get their head around climate change. Give it a bit of time to settle in. If you're having trouble with it, I'd really recommend reading *Carbon Detox* by George Marshall, or find a friendly climate campaigner with whom you can talk things through. Cutting down your own climate change impact can also help – try driving less, buying things second-hand, making your home more energy-efficient. Although these things aren't going to make the big social changes we need, they're a good place to start and can help you feel more in control and positive about things.

Take heart. We can do this. There are loads of people doing great stuff on this already, it's just that a lot of them are outside mainstream politics and so the media usually ignore them. Your involvement, however small, will help to move things in the right direction. This is your chance to be part of something really important.

Pick your passion. Look at the list below and find a topic or two that you'd like to do something about. Is there something on the list that particularly catches your interest, or makes you feel inspired, or angry, or hopeful? Is there an area where you already have a bit of knowledge, or that's relevant to your work, or studies, or hobbies, or your local community? Don't worry if not – there are groups and networks working on all these issues who can help you to get started.

Don't go it alone. Think about people you know who might also be interested in this. Getting just one or two friends on board can make a huge difference. Then link up with groups and networks that are already working on your area of interest – there are some suggestions below. Find out what's going on and see how you can help. If you don't have much time to commit, that's OK – just do what you can for now, and look for opportunities to do more later. Once you find the right project or campaign that grabs your imagination, you'll find yourself making time for it. If you want to start up something new, brilliant – but make sure you begin by gathering a few supporters and sharing ideas, and take advice from people who've done that sort of thing before.

Doing something in your local area and something a bit more big-picture is often a good combination. Petitions and letter-writing have their place, but things are getting rather urgent; we need to be ready to make real nuisances of ourselves. Avoid groups who seem too cosy with polluters or obstructive politicians. Support other activists from afar if need be, with bits of online work or donations or supportive letters to the media. Find ways to help people who are doing the things you can't do yourself.

Ten top tips to save the climate

Ten not necessarily easy (but effective and rewarding) things to do to save the climate

1) **Build the movement**. We need as many people as possible to get switched on, clued up and active. Think about what you can do to spread the word. Could you arrange a meeting or event at your workplace or in your community? Organize a talk, or a film showing, or a meal or a party. Find a book you like about climate change and lend it to all your friends (or if it's this book, buy it for them, obviously). There are lots of resources and groups who can help you get the message out there. Start collecting the contact details of the people you meet who want to become active too. Is there anything you could work on together?

2) **Stop the worst stuff: get off the coal train, shut down the tar sands and end the biofuel boom**. These are three of the most important campaigns in the world right now. Is there a local action group near you?

3) **Get the alternatives rolling**. What's the sustainability situation where you live? What's happening in your community, or region, or country when it comes to renewable power, sustainable transport, organic farming, or energy-efficient homes? Is there a local group trying to get community projects off the ground? How sustainable is your workplace, school, college, or place of worship? Are there opportunities to pressure or shame your local or national government to put more sustainable solutions in place?

4) **Reclaim democracy and clean up politics**. There are lots of groups pushing for democratic reform and cleaner politics that you could get involved with. This is a great opportunity to make links with other non-climate campaigners. We need to bring a sense of urgency into the struggle for better politics. Some climate activists are developing their own forms of fairer decision-making within their networks, and

trying to create more democratic structures within their local communities. There's much to learn from Latin America, where several countries rebuilt their democracies from the ground up. Find out more, talk to people, see what you can get involved in.

5) **Fight the growth myth**. We need to get people talking about this, challenging the bizarre notion of endless GDP growth and spreading the word about the alternatives. Get clued up about it and make sure other climate campaigners know about it too. It seems to be an idea whose time has come, with more and more people taking it seriously and getting vocal about it. Find them and help them.

6) **Switch off the carbon tap**. We need to rein in polluting industries through tighter regulation and grassroots action. We also need to change the global rules that work in favor of destructive development – the WTO, World Bank, IMF and free trade deals like NAFTA need either dismantling or radical reform. False solutions such as carbon trading, carbon offsetting and the myth of 'clean coal' are all dangerous distractions in need of determined debunking. People who are already on the case would love your help.

7) **Stick a spanner in consumer culture**. We need to challenge the strange dominant culture of shopping, driving and flying ourselves into oblivion. This is an opportunity for all you creative types – we need art, music, creative writing and performance of all kinds that goes against the grain and exposes the absurdity of modern consumer culture. Art and entertainment can reach people in ways that facts and figures never can.

8) **Link to other local campaigns**. People all around us are standing up for workers' rights, migrants and refugees, better healthcare, an end to poverty, and many other issues. There are very few issues that aren't related to climate change in some way – they'll either be being exacerbated by the climate crisis, or will have the

same root causes in our failed politics and rampaging economies. Get involved in these campaigns, and try raising climate change where it's relevant. Encourage local climate groups to make these links too. Look for obvious common causes – for example, climate campaigners united with labor activists to try to save a wind turbine factory on Britain's Isle of Wight in 2009. We need to learn and share knowledge together and build alliances.

9) **Pick a fight**. Sadly, there's probably some sort of climate awfulness going on near you – an airport expansion, proposed oil refinery, new highway or mass tree-felling. There's probably someone sensible resisting it. Join the resistance. These kinds of campaigns are important in themselves, but are also a brilliant opportunity to expose the bigger problems in the political system. Use it as a platform to condemn the rotten politics, corporate lobbying, and lack of regulation that allow this sort of thing to happen. Challenge the mantra of endless growth and corporations' legal requirement for profits that are (almost certainly) driving this development. Try to beat the bad thing, but make it about more than just that one bad thing.

10) **Support struggles on the climate frontline**. Look for opportunities to link local campaigns to Southern social movements and struggles on the ground. Talk to your local campaign groups and see if they're working on anything that could be linked up to a grassroots struggle elsewhere in the world. Does a local company or government office use products linked to the destruction of someone else's homeland? Why doesn't your local group get in contact with the community involved and find out what they'd like you to do about it? What can you do to build support for initiatives like the Save Yasuní campaign to keep Ecuador's oil in the ground?

How to talk to people about climate change

The Climate Change Communication Advisory Group (CCCAG), a UK network of climate communication experts, gives the following advice for talking convincingly to people about climate change:[1]

• Be honest and forthright about the probable impacts of climate change, and the scale of the challenge we confront in avoiding these, but avoid deliberate attempts to provoke fear or guilt.

• Be honest and forthright about the impacts of mitigating and adapting to climate change for current lifestyles, and the 'loss' — as well as the benefits — that these will entail. Narratives that focus exclusively on the 'up-side' of climate solutions are likely to be unconvincing. Avoid emphasis upon painless, easy steps. Avoid over-emphasis on the economic opportunities that mitigating, and adapting to, climate change may provide, or on the opportunities of 'green consumerism' as a response to climate change.

• Empathize with the emotional responses that will be engendered by a forthright presentation of the probable impacts of climate change.

• Make pro-environmental behavior seem normal, and harness the power of social networks.

• Think about the language you use, but don't rely on language alone.

• Encourage public demonstrations of frustration at the limited pace of government action. ∎

1 www.pirc.info/projects/cccag

Sustaining our energy

Keep an open mind. Be prepared to make unexpected allies and work outside your comfort zone. You might find yourself designing leaflets, attending local council meetings, involving your whole family in a community orchard, making sandwiches for people chained to a power station, or being interviewed on the national news. You may decide you want to be chained to the power station yourself next time, or you might prefer to organize a fundraising event to help pay for their court fees. Whatever your interest, and however far you want to go, there are loads of exciting and important things for you to get involved with.

If governments won't phase out fossil fuels, then we'll have to start doing it for them, by shutting down their coal mines and oilfields. If they won't protect the world's forests – or worse, if they try to sell them off for private profit – then we'll unite with the people of those lands and defend them ourselves. Let's set our own schedule for the closure of every coal mine, for the shutdown of the last tar sands pipeline, for the ultimate death rattle of the carbon markets. Let's lay out plans for the reclamation of indigenous peoples' lands, and for gaining community control over food, water and energy. Let's start to reclaim our world democratically from below, based on the growth of human rights, health and freedom, not the destructive fiction of economic growth.

Who knows how far we could get, how many positive changes we could make happen through our own actions? At the same time, we'll be laying down a powerful challenge to spur governments into action.

There's much to do, and we need lots of different people from different backgrounds to work together on this – including you. The good news is that at the Copenhagen protests and the Cochabamba talks we learned that it is possible to put our political differences aside and unite behind a common purpose – and that diversity really can be strength. We can work successfully together without having to agree on everything, so long as we're honest about our differences and create spaces in which to debate the issues openly. We have more than enough common ground to build a powerful movement. We'll find out who was right about all the details later on, after we've won.

We're waiting for you to join us. What are you waiting for?

Bibliography and Useful Contacts

Climate science and impacts (Chapters 1-3)

Global Humanitarian Forum, *The Anatomy of a Silent Crisis*, 2009.
James Hansen, *Storms of my Grandchildren*, Bloomsbury, 2009.
IPCC Fourth Assessment Report (2007), available at www.ipcc.ch
Elizabeth Kolbert, *Field Notes from a Catastrophe*, Bloomsbury, 2006.
J Timmons Roberts and Bradley Parks (2007) *A Climate of Injustice*, MIT Press, 2007.
Jerry Silver, *Global Warming and Climate Change Demystified*, McGraw Hill, 2008.

These websites are highly recommended for the latest climate science:

www.skepticalscience.com
www.realclimate.org

Greenhouse gas emissions (Chapters 4-5)

www.earth-policy.org
www.wri.org
www.noaa.gov
www.pbl.nl/en/index.html

See also the UK carbon emissions calculator that I built for the *Guardian* website:
www.guardian.co.uk/environment/interactive/2010/apr/21/national-carbon-calculator

Climate-friendly technologies (Chapter 6)

Accra Caucus, *Realizing rights, protecting forests: An Alternative Vision for Reducing Deforestation*, 2010. Available at http://nin.tl/9PV86b
Environmental Change Institute, *The 40% House*, University of Oxford, 2005.
GRAIN, Seedling special issue, 2009, available at http://nin.tl/ciRh9q
Mark Z Jacobson & MA Delucci, *Evaluating the Feasibility of a Large-Scale Wind, Water, and Sun Energy Infrastructure*, Stanford University, 2009.
David Mackay, 'Sustainable Energy Without the Hot Air', UIT, 2008, www.withouthotair.com
Dr Joe Romm's Climate Progress blog (http://climateprogress.org)
Zero Carbon World website at www.zcb2030.org/index.php/zcb-world

Bibliography and Useful Contacts

Climate history and politics (Chapters 7-10)

Joel Bakan, *The Corporation*, Free Press, 2004.
Climate Outreach and Information Network, 'Climate Change Condensed' training pack, 2010, available via www.coinet.org.uk
Wayne Ellwood, *The No-Nonsense Guide to Globalization*, third edition, New Internationalist, 2010.
Ross Gelbspan, *Boiling Point*, Basic Books, 2004.
Patrick Hossay, *Unsustainable*, Zed Books, 2006.
Robert Henson, *The Rough Guide to Climate Change*, Rough Guides, 2008.
Larry Lohmann, 'Carbon Trading', *Development Dialogue* no 26, Corner House, 2006.
Heather Rogers, *Green Gone Wrong*, Verso, 2010.
Derek Wall, *The No-Nonsense Guide to Green Politics*, New Internationalist, 2010.
Spencer Weart, *The Discovery of Global Warming*, book and website, 2009, www.aip.org/history/climate
CO_2 figures from National Oceanic and Atmospheric Administration and Carbon Dioxide Information Analysis Center.

www.sourcewatch.org
www.desmogblog.com
www.medialens.org

Taking Action (Chapter 11)

Help with spreading the word
The Climate Outreach and Information Network (www.coinet.org.uk)
Center for Climate Change Communication (www.climatechangecommuni-cation.org).
George Marshall, *Carbon Detox*, Octopus, 2007.

Agrofuels
www.biofuelwatch.org.uk
www.foe.org.au

Oil
www.oilwatch.org
www.priceofoil.org
http://planb.org

Coal
www.coalaction.org.uk
www.cana.net.au
www.coal-is-dirty.com

Tar Sands
www.ienearth.org/tarsands.html
www.no-tar-sands.org.
www.oilsandstruth.org

Cultural challenges
www.nationalpetroleumgallery.org.uk
www.museumofthecorporation.org
www.remembersarowiwa.com
www.artnotoil.org.uk
www.adbusters.org

Food and forests
www.viacampesina.org
www.wrm.org.uk
http://ran.org
www.sinkswatch.org

Sustainable alternatives
www.transitionnetwork.org
www.greennewdealgroup.org
www.neweconomics.org

Building a global movement
www.climate-justice-action.org
www.climate-justice-now.org
www.pwccc.wordpress.com
http://peoplesclimatemovement.net
www.globalclimatecampaign.org
www.globaljusticeecology.org
www.ienearth.org
www.climateconvergence.org

Australian action networks
www.risingtide.org.au
www.climatecamp.org.au
www.cana.net.au
www.climatemovement.org.au
www.foe.org.au

Canadian action networks
www.energyaction.net
www.uncampement.net
www.climateactionnetwork.ca
www.greenpeace.ca

Bibliography and Useful Contacts

New Zealand /Aotearoa action networks
www.savehappyvalley.org.nz
www.climatecamp.org.nz

UK action networks
www.climatecamp.org.uk
www.campaigncc.org
www.risingtide.org.uk
www.climaterush.co.uk
www.planestupid.org
www.wdm.org.uk
www.nonewcoal.org.uk
www.christian-aid.org.uk
www.foe.co.uk
www.greenpeace.co.uk

US action networks
www.actforclimatejustice.org
www.350.org
www.energyactioncoalition.org
www.risingtidenorthamerica.org
www.energyjustice.net
www.climateconvergence.org

Index

Index

Index

Index